THE SPATIAL WEB

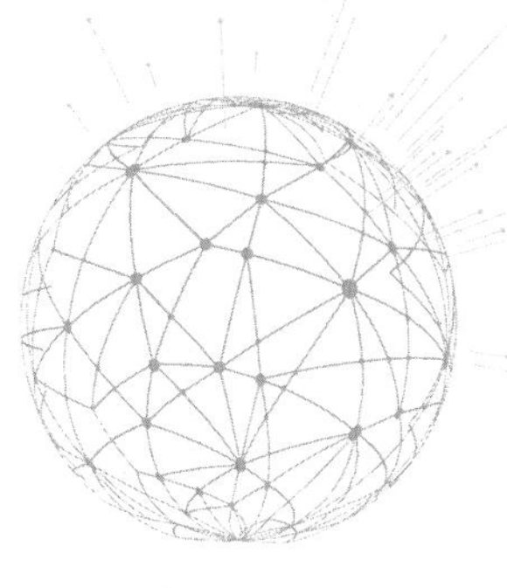

HOW WEB 3.0 WILL CONNECT HUMANS, MACHINES AND AI TO TRANSFORM THE WORLD

GABRIEL RENÉ + DAN MAPES

This book is dedicated to all future generations.

CONTENTS

THE SPATIAL WEB

PROBLEMS

FOREWORD

By Jay Samit, Former Independent Vice President Deloitte
Author of *Disrupt You!*

Don't you just hate it when you can't find your keys? One moment life is fine and the next you are frustratingly turning your house upside down like Indiana Jones searching for the lost ark. Americans spend 2.5 days per year just looking for misplaced items (where is that remote control?) and $2.7 billion dollars a year replacing lost necessities that remain missing (I now buy my reading glasses by the dozen).

Turns out humans are exceptionally good at losing things. Airlines annually lose over 3.5 million pieces of luggage. Cargo ships lose over 10,000 shipping containers per year valued at over $50 billion. Keys, glasses, and cargo can all be easily replaced. What happens when we lose things as valuable as knowledge?

The University of Alabama was created in 1820 with the esteemed goal of being a world-renowned center of learning. By the time of the Civil War, the University's library had one of the largest collections of books in America. When it was burned to the ground, only one book survived. Again, humanity is really adept at losing things. The world's greatest knowledge repository in 331 BCE, the Library of Alexandria in Egypt which had hundreds of thousands of Greek scrolls, is lost to history. Twelfth-century Turks burned the 700-year-old Nalanda library in India to the ground. One hundred years later, Mongols destroyed the Abbasid Caliphates House of Wisdom in Baghdad. In the New World, Spanish conquistadors lost the Maya Codices. The loss of priceless knowledge that takes centuries to relearn shortens lives and thwarts progress.

While losing scientific knowledge can set civilizations back centuries, it pales in comparison with the loss of human identity. Perhaps the most existential question of all, Who am I? is getting harder to prove in the 21st century. Far greater than the $16 billion in fraud costs annually caused by identity theft, millions of people are no longer able to prove who they are, where they came from, or what they own. Wars and regime change have displaced more than 68 million people around the world.

When government records are destroyed, the title and ownership of land or assets get lost and become unable to be substantiated. Courts are still trying to settle ownership claims to properties and artworks stolen in the Second World War. Today, there are now more refugees than after World War II and the numbers of stateless people are about to grow exponentially. When climate change floods the Ganges, Mekong, and Nile River Delta, another 235 million people will be displaced. The United Nations estimates an astonishing 1 billion displaced persons globally by 2050! But thankfully, technology can now save humanity from our losing streak.

Lost cargo, inaccessible knowledge, and corrupted data are global issues facing every business, nation, and economy. The Spatial Web not only easily solves all these problems, but provides new insights and data to drive the fourth transformation of computing: connecting the digital and physical worlds into one integrated universe of objects and ideas. The impact of this new Spatial Web will dwarf that of the Internet and change how we live, work, and thrive.

A mere couple of decades ago, the personal computer ushered in the first digital transformation by connecting humanity to an intelligent machine. The Internet powered the second digital wave by connecting individuals with all sources of knowledge. Mobile further expanded these connections by connecting people with billions of other people. As transformational as these three technologies were in changing how we live, they all were still restricted to functioning in a two-dimensional digital plane.

By leveraging the incredible data speeds of 5G (downloading a 2-hour movie in 3.6 seconds), and utilizing the power of edge computing, people will be able to combine the real world with useful data and wearables such as augmented reality glasses. Add in a layer of artificial intelligence drawing from real-time data from a trillion Internet-of-Things sensors, and our lives will go from googling for knowledge to our environment anticipating our needs.

Your smartwatch, which monitors your vital signs and compares them to all others in your cohort, can tell your autonomous car to take you to the hospital before you even realize you are having a heart attack. Your doctor will also be simultaneously notified and can provide customized instructions to the emergency room staff. Smart cities can prioritize and move traffic out of your path to assure the quickest travel time. Every medical procedure you undergo at the hospital will be immutably recorded onto your blockchain-based medical history profile. The Spatial Web will provide better patient care outcomes while cutting US healthcare costs in half. In fact, 85% of present healthcare costs are caused by heart disease and diabetes which AI and wearables are better suited for managing patient preventative maintenance. The Spatial Web holds the key to fewer doctor visits, fewer medical tests and procedures, and a lower demand for prescription medications. Healthcare is just a small example of the power of spatial computing.

Digital supply chains will interface seamlessly with sensors in warehouses and on retail shelves. Store shelves will place orders to warehouses when inventory is low. Just-in-time manufacturing will enable bespoke products delivered to your door. Maintenance crews can follow virtual arrows through buildings to find equipment in need of repair. Every mannequin in the department store will match your body dimensions and showcase the latest fashions accessorized by the clothing items you have previously purchased. Homebuyers can walk through potential homes virtually and see how their existing furniture fits the new home while spatially previewing carpets and window treatments from vendors that have the exact measurements of each room or window. Digital goods can be licensed to specific geographic locations and AI smart contracts will be able to make micropayments for a host of new goods and services. Trillions of dollars worth of new companies and innovative products will enhance our daily lives.

The Spatial Web holds so much promise for improving our lives that those companies failing to invest and embrace in the future will go the way of Kodak and Blockbuster. Companies that I work with such as Google, Amazon, Facebook, and Apple are already investing billions to provide the tools and platforms needed to utilize the Spatial Web. It is up to the entrepreneur to create the next generation spatial apps and services that will fuel the future. With this book, Gabriel Rene and Dan Mapes are the Lewis and Clark of this pioneering adventure. And as you are reading this, you are already in possession of the map they provided for your own personal discovery.

PREFACE

FROM FLATLAND TO SPACELAND

In 1884 English schoolmaster Edwin A. Abbott wrote a satirical novella about a fictional two-dimensional world of Flatland and a particular Square that encounters for the first time a three-dimensional entity known as the Sphere and its tales of Spaceland. This now-famous story has been told and retold for more than a century. Carl Sagan re-introduced the Flatland narrative into popular culture in his monumental Cosmos series. In the story, paraphrasing Sagan, the inhabitants of Flatland have width and length, but no height. Some are squares; some are triangles; some have more complex shapes. They scurry about, in and out of their flat buildings, occupied with their flat businesses and flat interests. They are familiar with left-right and forward-back but have no concept about up-down. Now imagine the inhabitants of Flatland. Someone may suggest that they imagine another dimension. And they respond, "What are you talking about? There are only two dimensions. Point to that third dimension. Where is it?" To them, the mere suggestion of other dimensions appears absurd.

As Sagan tells the story, "One day a three-dimensional creature—a sphere—comes upon Flatland, hovering above it. Observing a particularly attractive and congenial-looking square entering its flat house, the sphere decides, in a gesture of interdimensional amity, to say hello. 'How are you?' asks the visitor from the third dimension. 'I am a visitor from the third dimension.'

The wretched square looks about his closed house and sees no one. What is worse, to him it appears that the greeting, entering from above, is emanating from his own flat body, a voice from within. A little insanity, he perhaps reminds himself gamely, runs in the family. Exasperated at being judged a psychological aberration and in order to allow Square to be able to see the truth, Sphere descends into Flatland.

Now a three-dimensional creature can exist in Flatland, only partially; only a cross-section can be seen, only the points of contact with the plane surface of Flatland. A sphere moving through Flatland would appear first as a point and then as progressively larger, roughly circular slices. The square sees a point appearing in a closed room in his two-dimensional world and slowly growing into a near circle. A creature of strange and changing shape has appeared from nowhere.

Rebuffed, unhappy at the obtuseness of the very flat, Sphere bumps Square sending him aloft, fluttering and spinning into that mysterious third dimension. At first, Square can make no sense of what is happening; it is utterly outside his experience. But eventually, he realizes that he is viewing Flatland from a peculiar vantage point: 'above'. He can see into closed rooms. He can see into his flat fellows. He is viewing his universe from a unique and devastating perspective. Traveling through another dimension provides, as an incidental benefit, a kind of X-ray vision.

Eventually, like a falling leaf, Square slowly descends to the surface. From the point of view of his fellow Flatlanders, he has unaccountably disappeared from a closed room and then distressingly materialized from nowhere. 'For heaven's sake,' they say, 'what's happened to you?' 'I think,' he finds himself replying, 'I was "up."'

Sphere gave Square the ability to participate in interdimensional contemplations—to let us know that we need not be restricted to two dimensions. We can, as Sagan suggests, imagine higher dimensions.

In the grand arc of the human story, this book in a role similar to Sphere, presents ideas and forms from the third-dimension and beyond, transforming and translating into the language of Flatland—words. If effective these words will not only succeed in their aim to express or explain a new way of experiencing the world but also will inspire in the mind a more nuanced and multi-dimensional vision for how we might experience reality in the future. Like the Sphere, we hope to offer a brand new perspective to shift your frame of reference—to journey beyond Flatland, to help us to add new dimensions to the web, our world, our communication, and our very reality. Welcome to Spaceland.

PROLOGUE

"Study the past if you would define the future."

Confucius

n 1962, J.C.R. Licklider, the first Director of Information for the Pentagon's Advanced Research Projects Agency (ARPA), was struck with an extraordinary vision for the future. It was a vision of a new type of decentralized global computer network. Licklider believed that this new network would allow ordinary people, from anywhere in the world, to search digital libraries, communicate together, share media, participate in cultural activities, watch sports and entertainment, and make purchases of any kind by accessing computers from their own homes. He described it as *an electronic commons open to all, the main and essential medium of informational interaction for governments, institutions, corporations, and individuals.* He called it the "Intergalactic Computer Network."

In 1969, amidst the nuclear fears of the Cold War, the first phase of Licklider's vision was funded and the "ARPANET" was born. It was a new approach to a network, one that could protect against the kind of "single-strike attack" capable of delivering a lethal blow to America's then centralized telecommunication network. ARPANET achieved this goal with a new invention—packet-switching. Packet switching routed messages in the form of data "packets" between a decentralized network of computers (nodes) by allowing the packet to find its optimal route between the sender and receiver. This enabled a message to reach its ultimate destination even if one or more nodes in the network were attacked, compromised, or even destroyed, leading to the invention of the Internet Protocol Suite (TCP/IP). As more nodes joined, the more decentralized, secure, and incredibly valuable this interconnected network-of-networks or "Internet" became.

The nodes of the Internet were initially defined exclusively as "computers" and more specifically as computer servers—each with its own ID called an Internet Protocol Address or IP Address. However, the Internet has grown radically from those first four nodes in 1969, evolving through the eras of Web 1.0 (read-only websites on pc's) and 2.0 (social media on smartphones), and is currently on track to surpass 50 billion worldwide nodes. Those nodes have come to include our laptops, smartphones, watches, home appliances, drones, vehicles, and robots and one day, even us.

The first four nodes of the Internet circa 1969

Today as the world enters the era of Web 3.0, the power of the Internet and its decentralized design ethos will continue to extend into every aspect of our lives. We are about to add a trillion new sensors, beacons, and devices to the Internet of Things (IoT) over the next decade including exotic new types of wearable and biotech ingestible devices. This process will continue until we have computerized and connected every person, place, and thing in the physical world along with countless virtual objects and spaces. If it isn't obvious yet, the Web 3.0 era is about the **Internet of Everything**.

However…as the Internet transitioned from Web 1.0 to 2.0, we lost much of the distributed spirit of the original design principles. Corporations and governments are capitalizing on the surveillance, centralization, and monetization of web users and their data, in part, due to the limitations of the Web's inherent architecture. As we transition to Web 3.0, we have an opportunity to address these limitations and re-establish the decentralized nature that is central to the original vision— the creation of a global electronic commons, open to all. This is the Spatial Web.

INTRODUCTION

As we bear witness to the Digital Transformation of our world and cross the threshold into the Web 3.0 era, we face some extraordinary choices with serious and wide-ranging implications. Our technologies, from the first use of fire to the future of facial recognition, appear to be neutral by their very nature. Their appearance obscures their innate potential to magnify both the best and worst of human desires. Like the tale of Prometheus who stole fire from the gods to give to humankind to warm our homes and illuminate our civilizations, we must always remember that the gift of fire used to cook our food can just as easily become a curse that burns down our home.

The 21st-century technologies introduced in this book also contain the power to burn both ways, albeit at a previously unimaginable level of power and scale. And it is precisely for this reason that we as a species must carefully consider their use because the choices we make will fundamentally impact the lives of billions and set the stage for decades, perhaps centuries, to come. Our choices will not only determine the territorial lines of the web and the world, but also our very definition of the words humanity, civilization, and even reality itself. We must choose wisely.

The emergence of smart cities and factories, autonomous cars and homes, smart appliances and virtual reality worlds, automated shopping, and digitized personal medicine are transforming the way we live, play, work, travel, and shop. All across the planet, our technologies are breaking out from behind the screen and into the physical world around us. Simultaneously, the people, places, and things in our world are being digitized and brought into the virtual world, becoming part of the digital domain. We are digitizing the physical and "physicalizing" the digital. Clear boundaries between the real and the virtual are dissolving. Our near-term future has all of the indicators that the technologies that we've seen in our science fiction stories over the last century will be realized. Looking back a century from today, will we consider our sci-fi stories prophetic tragedies that we ignored or cautionary tales that we heeded?

Across the many "Top Technologies for 2019" lists compiled by the leading research firms around the world, a recurring pattern emerges. The investments and acquisitions that the largest tech companies have made over the last decade in Artificial Intelligence, Augmented and Virtual Reality, Internet of Things including smart cars, drones, robotics, and biometric wearables as well as Blockchain, Cryptocurrency, 5G Networks, 3D printing, Synthetic Biology, Edge, Mesh, Fog Computing, are being woven into the fabric of our global civilization, under our feet, above our heads, and all around us.

These technologies, all coming of age in the 21st century are often referred to as "exponential"—true to Moore's Law, about every 18 months their capabilities and performance powers double for the same cost. This is a common effect of computing technologies, known as "Moore's Law" for the co-founder of Intel, Gordon Moore.

Moore's Law is about compute power. And it is an often-quoted law of computing.

Robert Metcalfe was the founder of 3Com and one of the inventors of Ethernet. Metcalfe's Law states that the total value of a network

Moore's Law (Transistors per Microprocessor), 1971-2014

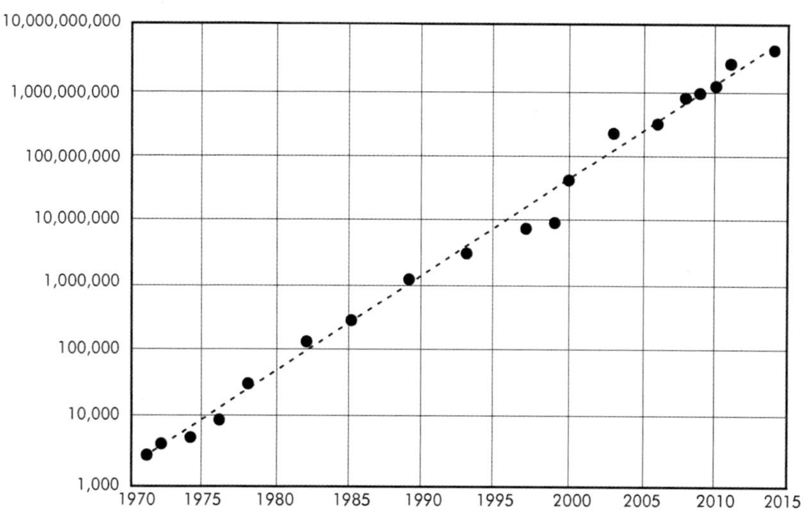

is determined by the number of other users connected to the network. The more users that an individual user can reach through a network, the more valuable the network becomes, even when the features and price of a service remain the same. Metcalfe's Law is about network _scale_.

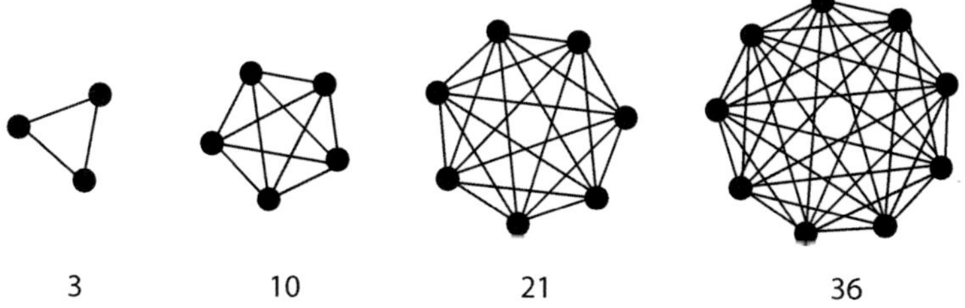

Although not every user or node is actually of equal value in any network, according to a recent study by the NfX Group titled "70 Percent of Value in Tech is Driven by Network Effects," over the last 25 years, approximately 70% of the value created in the Technology sector has been the result of companies that use these "network effects" to grow.

Arguably, the key reason we have seen Web 2.0 tech companies and start-ups become the largest companies in the world in terms of market cap, profit margins, and number of users is because they are the best examples of Moore's Law supply-side cost reductions in *computing power* complemented by the demand-side economies of *network scale* and value, as defined by Metcalfe's Law.

As an example, an upgraded smartphone on a faster network enables more engagement, content consumption, and sharing, which drives other users to participate. This has been the winning formula in the Web 2.0 era, which has driven adoption from 1 billion Internet users in 2005 to nearly 4 billion in 2019. 3.9 billion people use mobile phones to access the Web, and 2.5 billion of those use smartphones. 3.4 billion are on social networks, 3.2 billion of whom use their mobile phones to access social networks. See the trend? Better hardware, faster networks, more users. Exponential compute power + Exponential network connections = Exponential Value

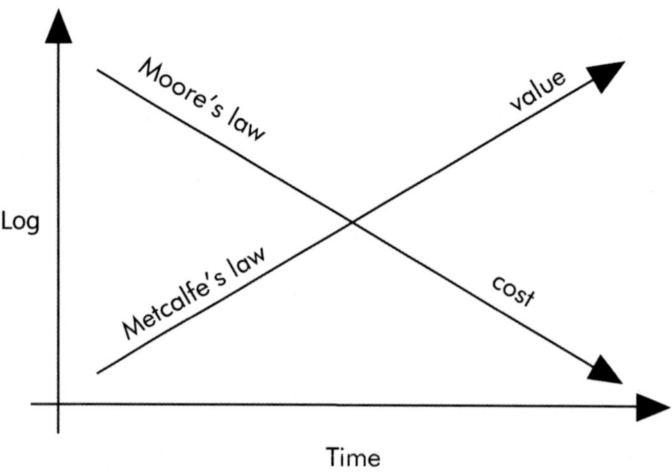

Just as computing power has increased even as the cost and size have shrunk from room-scale to desktop to laptop to handheld, our computer networks have grown from room-size, to company-wide, to country-wide, to worldwide networks.

Simply put, Moore's Law makes computing faster and cheaper while Metcalfe's Law grows the size and value of a network.

It will soon become technologically and economically worthwhile to computerize complex things like cars, cities, and humans, but ultimately, we will computerize increasingly smaller things like jewelry, clothing, buttons, and even our very cells.

And because value is best accumulated by combining many networks into a single network as the existing Internet has done, this new network of people, places, and things, rules, and value creates the Internet of Everything in the Web 3.0 era because it grants the ability to _computerize and connect_ everything.

This is the great "Convergence" of all computing technologies and all networks. The implications of this Convergence are unprecedented in power, scope, and scale. Fundamentally one must ask: What happens when we integrate the physical, digital, and biological domains together?

The Convergence

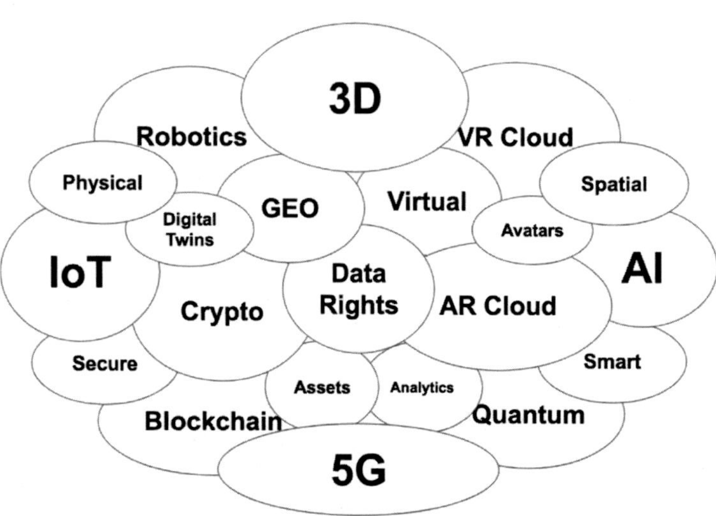

Will this Convergence lead us into the futuristic dystopia of our nightmares—the one where we are jacked into a simulated reality, a siloed and centralized network where we are constantly under surveillance by dark forces operating in the shadows, all the while being mined for our data like a natural resource? In Web 3.0, will we move from the Social Web to the *Shadow Web*?

If the past is an indicator of the future, then this is a real possibility. However, it is one that we should avoid at all costs. And we can, because the Convergence does, thankfully, grant us the opportunity to use these same technologies for amazing good: to address and even resolve some of civilization's biggest challenges; to turn us from a dark and increasingly bleak forecast towards a brighter and better future. Such a future could positively transform every aspect of our lives—how we work, learn, play, celebrate, and enjoy life. It could grant us the means for generating a prosperous global economy, thriving planetary ecology, and a healthy, advanced civilization. Facilitated by new network protocols, the Convergence could lead to the creation of a new web that connects physical places together with virtual places. It could enable an open and interoperable new generation of the web—a Web 3.0 era that secures the privacy and property rights of individuals while ensuring secure and trustworthy interactions and transactions between the human, machine, and virtual economies. This future literally adds a new dimension to the web. It enables —The Spatial Web.

The Spatial Web weaves together all of the digital and physical strands of our future world into the fabric of a new universe where next-gen computing technologies generate a unified reality; where our digital and physical lives become one. This is a new kind of network, not merely one of interconnected computers like the original Internet or a network of interlinked pages, text, and media like the World Wide Web but rather a "living network" made up of the interconnections between people, places, and things, their virtual counterparts, and the interactions, transactions and transportation between them. Like the

The Spatial Web

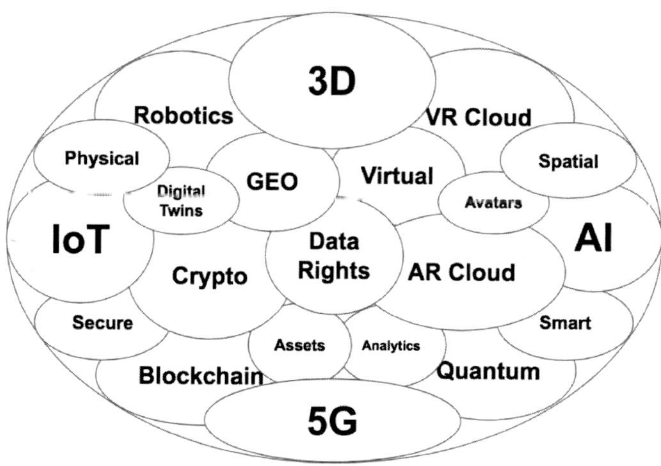

World Wide Web before it, this new Spatial Web requires new code to bring it to life. Not merely software code, but critically, ethical, and social codes as well.

Today, we can't share knowledge all that effectively because we cannot share our mental models and maps of the world directly between brains. We can't copy/paste ideas and concepts between ourselves, much less create, edit, or share them with AI, or IoT devices.

The human mind constructs a 3D mental model of reality that (today) cannot not be directly shared between others. Whereas we think and reason spatially, traditionally, we've relied on language, words, and visual representation in the form of 2D or 3D drawings to convey spatial information and context to others. The capture, transmission, and interpretation of objects in space acts is like a Private Virtual Reality (VR) that must be translated into a foreign tongue of words, text, or pictures in order to be shared. We're forced to use 2nd order mediums that reduce and degrade our internal models. Along the way, we lose much of the fidelity, nuance, and context.

But what if we had the technology to collaborate with machines and artificial intelligence to share collective 3D models and maps of our experience and knowledge directly? Instead of being locked in our Private VR we could share in a kind of multi-dimensional Public AR—a digitally-mediated consensus reality — an idea as exciting as it is concerning. However, using Distributed Ledger technologies like Blockchain could enable us to reliably verify factual data about the world while enabling the freedom to collaboratively explore, edit, mashup, and remix reality. These technologies could help us confirm the difference between the real and the "projected," between the empirically valid and the creatively expressed, providing us an Augmented Intelligent Reality.

The power of the Spatial Web initially comes from its ability to describe the world in the language that the world speaks to us in—geometry. The Spatial Web lets us use a digitally-mediated universal language in which all information can become spatial. It enables the current information on the web to be placed spatially and contextually on objects and at locations, where we can interact with information in the most natural and intuitive ways, by merely looking, speaking, gesturing, or even thinking. But it also enables the Web to be more physical as sensors and robotics become embedded into our environments and onto the objects around us. It makes our world smarter as it adds intelligence and context to any place, any object, and every person that we encounter, and it allows our relationships to each other and this new network to be more trustworthy, more secure, and faster by decentralizing and distributing the computing and storage of the data. It will enable us to accelerate and improve, augment and enhance every facet of our existence—our education, our creativity, our health, our businesses, our legal system, our politics, and our ecologies. The Spatial Web has the potential to move us from predominantly egocentric and ethnocentric concerns to more worldcentric ones that are more holistic, equitable, and inclusive.

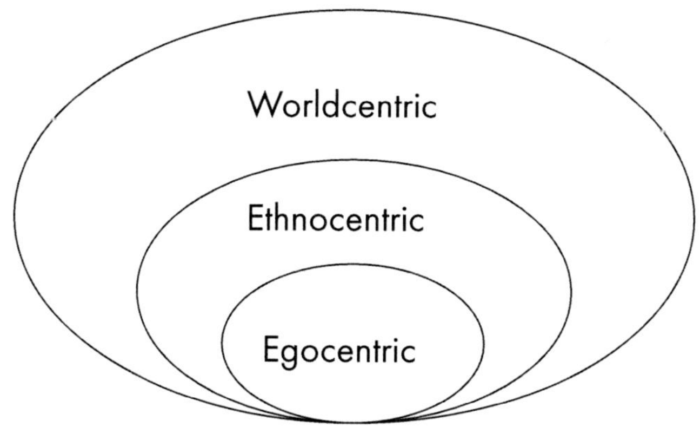

Despite the promise of a new Spatial Web, today we are still using web technologies invented decades ago that have significant limitations with regard to their use in the physical world. Today's web protocols were designed for interconnecting pages on computers, not people, places, and things in the real world. These protocols were designed for sharing information, not for managing and coordinating the activities of humans, machines and AI or engaging in real-time global trade and commerce.

Web 1.0 consisted of static documents and read-only data on PCs. Web 2.0 introduced user-generated multimedia content, interactive web applications, and social media on multi-touch smartphones. Web 3.0 marks the rise of AR and VR headsets, smart glasses, wearables, and sensors. These will enable the Spatial Web to project our information, ideas, and imaginations into the world around us, weave them into every conversation, displaying them in our cities, in the places we work, learn and live. With intuitively placed information, AI-assisted interaction, cryptographically secure information, and digital payments, a new kind of network is emerging. One where the Web becomes the World.

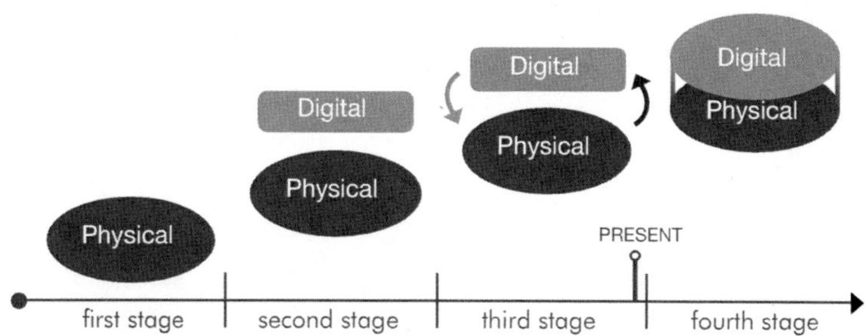

In Web 3.0, we will not only create a "Digital Twin" or soft copy of our world and everything in it, but a Smart Twin of everything, with its own unique ID, interaction rules, and verifiable history capable of being linked and synced to its physical counterpart, spatially. Like a Wikipedia for anything, making all the world searchable, allowing for any object, person, process, and system to be updated, quantified, optimized and shared—if you have the right permissions to do so. The applications are endless and the implications, both positive and negative—practically indescribable. However, it is important that we begin to describe them; and accurately, at that, if we are to benefit from its promise and avoid its pitfalls, we must first name the thing correctly.

Plenty of Augmented Reality or AR-centric names for a Digital Twin of the physical world have been put forth by various industry leaders. Augmented World Expo's founder Ori Inbar has suggested the term AR Cloud, a digital point cloud or mesh-like scaffolding of environments that allow us to project holograms into the world in ways that can be persistent and therefore experienced by multiple parties. Magic Leap has put forth the term *Magicverse*, a more playful version of a global digital twin that inspires visions of the fantastical. And in early 2019, acclaimed futurist Kevin Kelly of *Wired* magazine fame wrote an entire feature entitled *Mirrorworld* which attempted to consider a broader description that mentions IoT and other technologies in the Convergence but lacked a description of how these technologies would work together. Regardless of your preferred term, they are each meaningful attempts at the near impossible—describing an inevitable future world where the digital and the physical blend together, where the "world will be painted with data" as notable Forbes writer and industry XR veteran Charlie Fink is fond of saying.

What each of these terms lack is a clear vision for AR in the glaring light of the Convergence. How will this Digital Twin of the world combine with IoT, AI, and Blockchain to become a *Smart* Twin? How will AR work across various devices, operating systems, and locations? How will it work with other *realities* like Virtual Reality, the slower (projected adoption curve) but arguably larger cousin of AR? VR will play out its own exponential role—a 3D Virtual Reality internet of all connected VR spaces, games, and worlds classically referred to as the "Metaverse" in Neil Stephenson's seminal sci-fi novel *Snow Crash* and more recently described in Ernest Cline's book *Ready Player One* as "The Oasis." Whether we use these terms, the "Internet of Non-Things" or simply call it the "VR Cloud," we cannot casually omit it from the Convergence, banish it to its own corner of the Spatial Computing universe apart from AR. This is why a more comprehensive term is required, one that encapsulates all of the technologies of the Web 3.0 era and even provides the ability for users and digital objects to move seamlessly between AR and VR spaces.

Given the historical importance and exponential power ascribed to Convergence technologies, a comprehensive vision is required that describes how these technologies will be best aligned with our core human values and what the implications will be if they are not. Piecemeal descriptions and industry-centric narratives do not provide the holistic vantage point from which we must consider how best to make the critically important decisions regarding matters of privacy, security, interoperability, and trust in an age where powerful computing will literally surround us.

If we fail to make the right societal decisions now, as we are laying the digital infrastructure for the 21st century, a dystopic "Black Mirror" version of our future could become our everyday reality. A technological "lock-in" could occur, where dysfunctional and/or proprietary technologies become permanently embedded into the infrastructure of our global systems leaving us powerless to alter the course of their direction or ferocity of their speed. A Web 3.0 that continues its march toward centralized power and siloed platforms would not only have crippling effects on innovation, it would have chilling effects on our freedom of speech, freedom of thought, and basic human rights. This should be enough to compel us to take thoughtful but aggressive action to prevent such a lock-in from occurring at all costs.

Thankfully, there is also a "white mirror" version of Web 3.0, a positive future not well described in our sci-fi stories. It's the one where we intentionally and consciously harness the power of the Convergence and align it with our collective goals, values, and greatest ambitions as a species. In the "white mirror" version, we have the opportunity to use these technologies to assist us in working together more effectively to improve our ecologies, economies, and governance models, and leave the world better than the one we entered.

In the first quarter of 2019, an historically unprecedented but little-noticed event took place. The screens of over 1 billion new smartphones became windows into the physical world because both Apple and Android added AR software to their operating systems. This gave smartphones the capability to display 3D contextual information and 3D interactive objects in space. Apple, Microsoft, Google, Samsung, Facebook, Magic Leap, Baidu, Tencent, and others have also invested billions of dollars into a new generation of smart glasses and headsets for Augmented and Virtual Reality designed to initially supplement and ultimately replace our smartphones. In addition, billions of cameras and 3D depth sensors soon to be added to our existing smartphones and to the billions of drones, robots, cars, and street corners, will make them all spatially aware, capable of mapping their environments in 3D and acting as windows to the Spatial Web. These digital devices will enable us to make a complete copy or Digital Twin of the world we live in. Over the next decade, as we complete this transition, spatial technologies will become the leading interface, not only for our various digital activities but for our physical ones as well.

The following pages lay out the need for a Spatial Web and describe how the technologies of the Convergence will be connected by

The Internet of Things

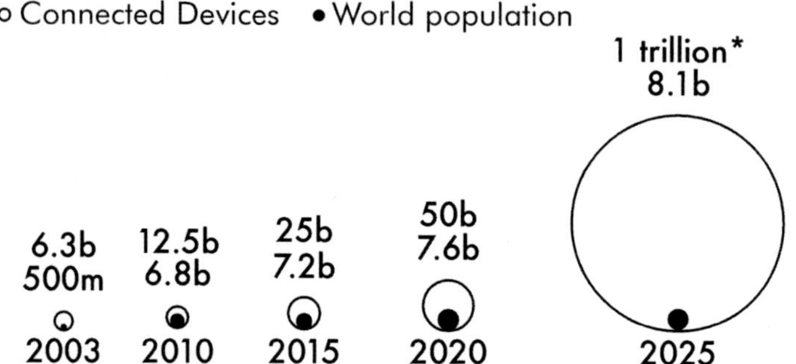

new software protocols will bring it into existence. They further outline the applications and implications for us as individuals and as a society. Our intention, as authors, is for this book to serve as a primer to acquaint you with each of these powerful technologies, and to lay out a vision for a path forward that inspires and invites you, the reader, to participate in the development of a more open and free new web… a new web that stands on the shoulders of those who built the earliest Internet and web infrastructure, but looks to our future, to lay down a new universal digital infrastructure that is designed to better align the power of our technologies with our human values.

The VERSES Foundation is a non-profit committed to free and open standards across emerging technologies. Our organization, with input from an extraordinary group of pioneering thinkers, has architected a theoretical and technological framework for a universal digital infrastructure for the 21st century. To bring this vision of Web 3.0 to life will take us all to advocate, develop, iterate, and realize it.

Welcome to The Spatial Web.

THE SPATIAL WEB

"This magical future ahead is called
the Spatial Web and will transform
every aspect of our lives."

Peter Diamandis

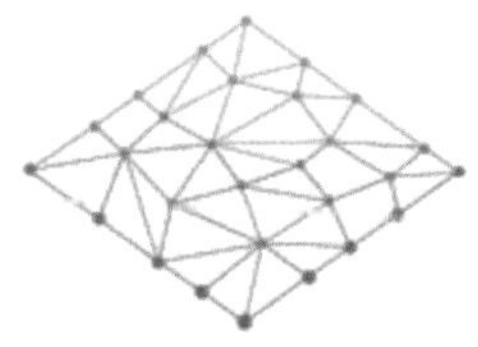

World Wide Web
websites linked together

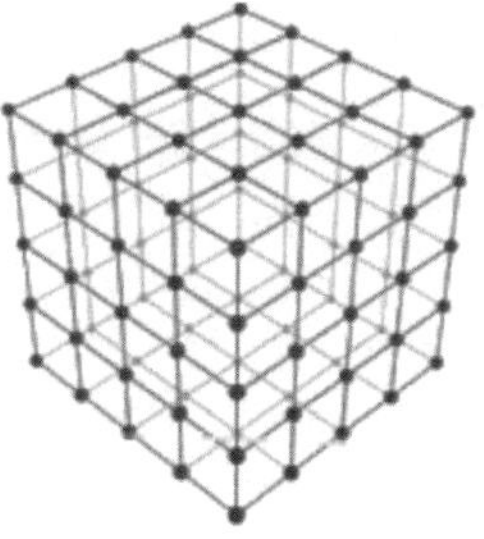

Spatial Web
people, places, and things
linked together

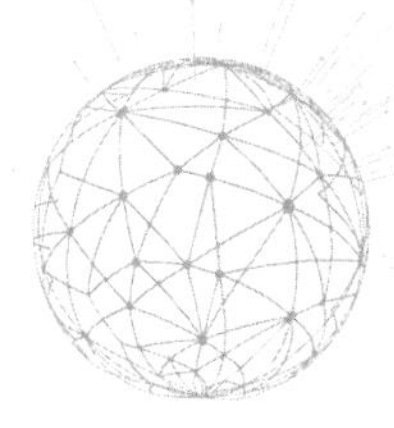

THE DREAM OF
THE SPATIAL WEB

The visions of the future imagined in our sci-fi films, TV shows, books and games depict worlds where advanced technologies project users into fantastically immersive, interactive experiences and allow them to engage with holographic content and characters via user interfaces that magically integrate the virtual and physical worlds. Movies like *Blade Runner*, *The Matrix*, *Star Wars*, *Avatar*, *Star Trek*, *Ready Player One* and *Avengers* show us futuristic worlds where holograms, intelligent robots, smart devices, virtual avatars, digital transactions, and universe-scale teleportation capabilities all work together perfectly within a kind of unified reality that somehow seamlessly combines the virtual and the physical with the mechanical and the biological. Science fiction has done an excellent job describing a vision of the future where the digital and physical merge naturally into one—in a way that just works everywhere, for everyone. However, none of these visionary fictional works go so far as to describe exactly how this would actually be accomplished.

But it has inspired many of us to ask the question—How do we enable science fantasy to become.... science fact?

We inhabit two different worlds. One is the physical world, governed by space, time, matter, and physical laws. This is the world where we eat, breathe, live, and work. In this physical world, food, water, and shelter are still the fundamental requirements needed to live.

The other is the digital world. It is not governed by space, time, matter or physical laws. This world serves as a canvas for our internal states, where we capture and share our thoughts, feelings, ideas, information, and imaginations with others. It's like an external mirror which we project our individual and collective interior selves upon, even as it reflects these projections back to us. These worlds exist apart from the other—separated by a pane of plastic or glass. Though connected by threads of information, they have remained functionally separate with no means of reconciling their distinct and dissonant perspectives of space, time, and physical/non-physical laws. As our physical reality and digital realities collide, the implications for humanity and our future, remain unknown.

It is possible that this new *mixed reality* may ultimately force us to rewrite our definition of reality itself as it becomes as collaborative, malleable, and editable as the print and electronic media of the 20th century became in the Web 2.0 era. Considering this possibility, how will we as a species ensure that we will be able to maintain the trust, security, privacy, and interoperability necessary to gain the greatest promise from this new reality while avoiding its most threatening pitfalls in the Web 3.0 era?

Let's begin by properly defining the term Web 3.0.

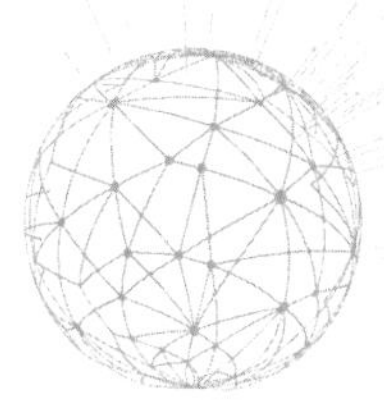

DEFINING WEB 3.0

Early definitions of Web 3.0 have predominantly defined it as the Semantic Web. The Semantic Web put forth the idea that the text we read on the web would become contextual—the intended meaning of words or sentences could be made explicit and therefore "semantic." Encoding the contextual meaning of a word into the text on a web page would enable both the text and eventually the Web itself to become "smart." The smart web would know, for example, that on a site dominated by discussions of how best to grow healthy plants and vegetables indoors, a particular reference to the term "greenhouse" would be a reference to a glass building filled with plants rather than a house painted a shade of green. A great idea to be sure, but sadly, it was simply too difficult to reverse engineer a powerful new layer of intelligence to the world wide web.

Beyond this technicality, what if the shortcoming for the vision of a Semantic Web isn't in the scale of its ambition, but rather in its limited focus? In Web 3.0, the domain of what can become smart, contextual, and consequently "semantic" will not be limited to text but extended into the physical world itself, where spatial objects, environments, and interactions dominate. Web 3.0 *will* be a semantic web, but not because we have embedded intelligence into text. It will be semantic because we will embed 3D spatial intelligence into everything.

Suffering from a case of industry-centric myopia, many contemporary definitions of Web 3.0 also lack holistic thinking. For example, Web 3.0 will not be limited to being a cryptocurrency-driven, peer-to-peer "Internet of Value" as many of the Blockchain faithful claim, or an "Internet of Intelligence" driven by a network of (hopefully) benevolent AI as the Artificial Intelligentsia suggest. It will not merely be the trillion device "Internet of Things" as Industry 4.0 advocates espouse or the "Internet of Me" where various wearables and ingestibles will be able to track every pulse, personalize every meal, and optimize every step, emotional state, and eventually even thought. It will not even be the long-prophesied 3D Internet of interconnected worlds, a Virtual Reality "Metaverse" or VR Cloud or its more recent counterpart, the AR Cloud and its "Internet of Places" as the Spatial Computing initiates may believe. No. Web 3.0 will not be limited to any one of these definitions because, in this next era of the web, it includes all of them. In Web 3.0, all of these get "connected" into the *Internet of Everything*.

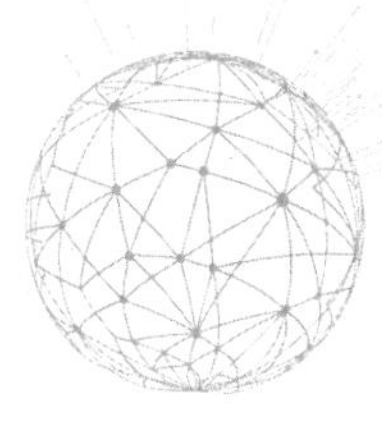

THE EARLY POWER OF SPATIALIZATION

The term "spatial" in the Spatial Web references how our future interfaces enable a web that extends beyond the screen to integrate and embed *spatial* content and interactions, facilitated by *distributed* computing, *decentralized* data, *ubiquitous* intelligence and *ambient, persistent,* edge computing. Each of these technology trends fundamentally extends computing power *further into the space around us,* bringing new dimensions of experience, connection, trust, and intelligence to the world. We call this macro-computing trend Spatialization.

The Spatial Web's imminent arrival can be glimpsed in the precursor startup "unicorns," like Uber, Airbnb, Snap, Niantic, Postmates, and TaskRabbit, many of which rapidly rose to valuations exceeding $1 billion. These startups owe their success to the hardware enhancements of the smartphone—in particular, the location-based functionality imparted by GPS, the positional and directional capabilities of gyroscopes and accelerometers, and the miniaturization of camera technologies. These technologies drove massive user adoption and multibillion-dollar valuations because they unlock the value of a key trend without which they would not even exist—spatialization.

For example, Uber rents a space that will transport you from one location to another. Airbnb rents a space that you can stay in when you get there.

Snap and its new Lens Studio, Tik Tok, YouCam, and others are parts of a fleet of new smart camera-based products that offer the ability to alter, morph, augment, or change the nature of your face by adding crowns, horns, adaptive facial features or detailed makeup applications, a superimposed character over your own head, or even the background and environment around you. Niantic's Pokemon Go exploded onto the scene with a global treasure hunt for unique anime characters that were strategically placed at various locations across the planet. A billion people walked out of their homes and chased virtual characters across the spaces in their towns, parks, shopping malls, and streets. The ability to have food delivered to your location via Postmates or order someone from TaskRabbit to mount a flat screen in your apartment is based on technology that navigates them to and from locations—a specific point in space, not on a screen.

It is the digitization of location and commercialization of spatial tasks that form the foundation for the Spatial Web.

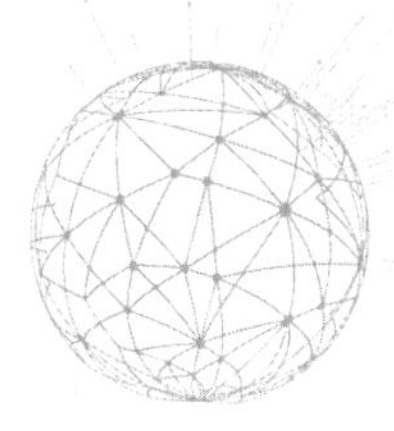

THE EVOLUTION
OF DIGITIZATION

Digitization is the process of converting the analogue world around us into a code made of zeros and ones (bits) so that our computers can read, store, process this information and transmit it across digital networks. Fundamentally, digitization disrupts and democratizes the production, storage, and distribution of whatever medium it transforms. However, whereas humans learn spatially first and symbolically second, our computer systems "learn" or "process" in the opposite order.

Computers began with digitizing symbols and are now gaining digitized spatial awareness and this drive towards the spatialization occurred over several stages. For example, computers first gained the capacity to digitize numbers. This led to advancements in mathematical computation that assisted us in data storage, statistical analysis, and code-breaking among other things. This phase of digitization essentially used computers as "math processors."

Next, computers became capable of digitizing text, and the "word processor" was born. Word processors allowed us to dynamically edit, save, and share text. Digital words powered the birth of new coding languages, desktop publishing, and email. The digitization of words also spawned the creation of "hypertext," which led to the World

Wide Web and Web 1.0 was born.

The digitization of media came next. The power of the smartphone enabled digital media to be captured and shared at an unbelievable scale and pace. Web 2.0 was built on digital media creation and consumption backed by the power of the network effect of the web which accelerated the adoption of the smartphone. The smartphone became the web's most efficient and effective "media processor."

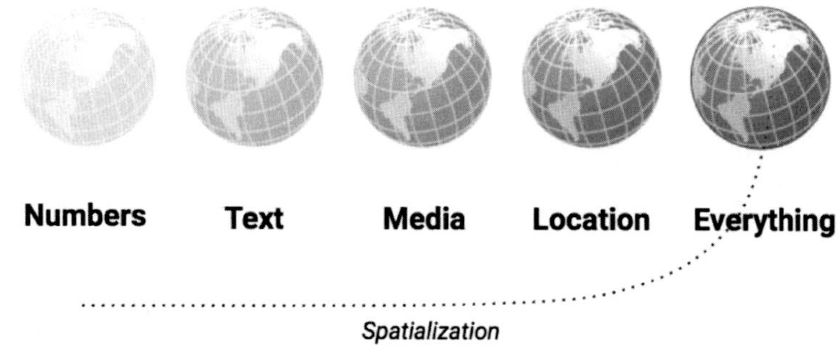

Spatialization

Using 3D modeling and animation tools, we have recently been digitizing "things" from every aspect of our lives across many different industries including television and motion pictures, video games, marketing, advertising, and augmented and virtual reality, but also for all of our modern product and industrial design, architecture, civil engineering, and city planning.

Today, 3D models can be produced by scanning real-world objects and locations using the next generation, depth-sensing computer vision powered cameras installed in smartphones, drones, and cars. These cameras have the ability to scan and build 3D models and maps of products, objects, humans, buildings, and even entire cities. Connected devices are capable of mapping and tracking our faces, fitness levels, movements, moods and health. Smart sensors are being embedded into manufacturing and industrial equipment to track and monitor speed, pressure, temperature, and much, much more.

So, we have progressed from the digitization of text and media and pages to the digitization of people, places, and things, not just digitizing our symbols and media, but the objects and activities of our physical reality. To complete our "computers as processors" metaphor, computers at this stage of Digital Transformation become "reality processors."

Fundamentally, digitization democratized the production, storage, and distribution of the medium it transformed at every stage. Looking back at the Impact our computer processors have had in digitizing our words and media; now imagine this impact when we apply it to every object, person, location and activity in the world.

Although the power of spatialization can boggle the mind, the real power of occurs when computers become networked together. However, the communication protocols that we've relied on in the past to link the words and pages of the symbolic realm of the World Wide Web are ill-equipped for the digitally-enhanced physical realities of the Spatial Web. This is because they are based on a very different information model for describing reality.

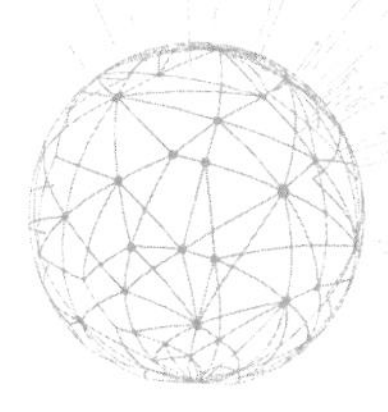

A NEW MODEL OF THE WORLD

Let's explore the difference. The way that we "model" our world, the ways in which we share it, the lenses that we look through, haven't fundamentally changed for thousands of years. The model that civilization has used to share information over time has predominantly been via "words on a page." A book.

The "page" has evolved a lot over time, becoming lighter and more easily portable. What began as stone or clay tablets, became papyrus scrolls before being bound together into a codex. These were later replaced by animal skin parchment which more clearly resembled pages, leading ultimately to the creation of the modern paper-based bound book.

The printing press—often considered humanity's greatest invention after fire—enabled publishing at a massive scale. This new "machine" popularized the method for copying something using a mechanical press that eventually could use plastic, polymers, and even metals to "press" copies of the various products that we use today.

Ever notice the "seam" on a plastic toy, appliance, or even car? That seam is there because nearly all the components of our products since the invention of the printing press are essentially "printed" as copies of various "sheets" that are assembled or "bound" together. It turns out that nearly everything is produced like a book. We make almost all products today using the mass-production techniques that were first invented for the mass-production of books.

The "movable type" of the printing press, first powered by hand, then steam, and then by electricity, ushered in the Industrial Age and its evolution from mechanization to electrification to computerization. Along the way, the digitization of words and numbers combined with the typewriter, calculator, and the microprocessor and the modern computer was born. The "killer app" for the first personal computers was word processing, the two most popular programs being Microsoft's "Word" and Apple's "MacWrite."

This brings us back to the World Wide Web. It is truly one of our greatest and most amazing modern inventions but it is also quite literally an offspring of the printed book. Even HTML or Hypertext Markup Language, the dominant language for editing web page content, is the combination of two printing abstractions—hypertext and what is called a "mark-up" language.

Mark-up languages have been in use for centuries. The idea and terminology evolved from the "marking up" of paper manuscripts with revision instructions by editors, traditionally written with a <u>blue pencil</u> on authors' manuscripts. For centuries, this task was done by skilled typographers known as "mark-up men" who marked up text to indicate what font, style, and size should be applied to each part before passing on the manuscript for typesetting by hand. One of the earliest digital mark-up languages that used tags to separate presentation instructions and content was called..wait for it... "Scribe."

Scribe primed Standard Generalized Markup Language (SGML), which became an International Standards Organization (ISO) standard. A few years later, a young Sir Tim Berners-Lee mashed up two elements—SGML and hypertext -to create Hypertext Markup Language (HTML).

He then combined HTML with a revolutionary new protocol that could connect these "hyperlinked" pages called Hypertext Transfer Protocol or HTTP, and a new web browser called NEXUS, and the World Wide Web was born in 1990. Marc Andreessen then created the more powerful and popular Mosaic browser in 1993 and that helped to spread the use of the Web around the world.

The World Wide Web is a global digital library. It uses hypertext to link between the pages of the websites (books) in the library. Just as we used to ask a librarian to help us find a book that we were searching for, today, we ask Google to help us find the page on the web (library). To really emphasize the metaphor, we even call the process of reading a page online, "scrolling." And Facebook? Well...it even has "book" in its name.

The written word and its consistent themes of words, pages, and publishing formats dominate the design architecture of the World Wide

MODEL OF THE WORLD

A Book

The World

Web, highlighting the idea that the way that we share digital information today remains dominated by the model of the "book." It has been the key to the cultural and scientific evolution of humanity, but at best, it is symbolic, two-dimensional information ***about the world***, trapped in pages, set behind a pane of glass. It is an abstraction layer. It is not the world itself. It was not designed to connect people, places, and things in the real world and It was not designed to include activities in the physical domain, i.e. to operate *spatially*.

Renowned scholar Alfred Korzybski's famous dictum "The map is not the territory" reminds us that our words only capture a small percentage of the actual world. But with the Spatial Web, the *map becomes the territory*, moving us beyond merely reading "about the world" in books or on screens, to engage the world directly where information is presented *in the world* and ***as the world***. It is time to evolve our model of the world from that of "the book," to a new model—*the world itself.*

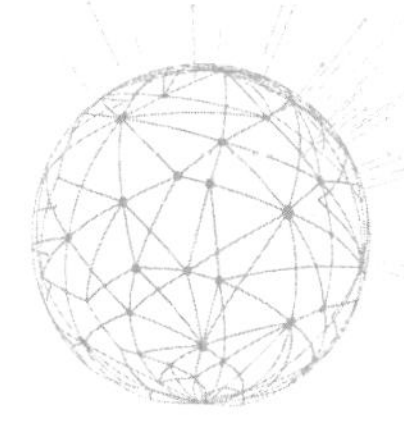

THE INEVITABILITY OF A SPATIAL INTERFACE

Human beings are spatial creatures and we live in a spatial reality.

Our biology has evolved for billions of years within a spatial environment. Our vision, hearing, cognition, and movements were all developed within the context of being spatial creatures occupying a spatial reality. We experience three spatial dimensions (six directions) plus time, and everything that we experience as "reality" is contained within these dimensions.

It could be argued that the dominant theme over the course of human history has been our impulse and desire to control our environment. Guide it with our hands, turn it to our will, and transform it into things that we find useful and meaningful. To extend our control over reality humans create technology. The earliest known technology was a "stick" our ancient ancestors used to pry termites from a mound in the ground. With this stick, they augmented their reach beyond their physical limitations to gain access to a rich source of protein.

Technology augments and extends the capabilities of the human body and brain. From the most primitive digging tools to the most advanced robotics, from the earliest abacus to the leading-edge artificial intelligence, our technologies have exponentially increased our ability to exert control over space, time, and matter (e.g., manipulate our environment for our collective benefit).

Digitization is simply the latest technology in a long line, invented to increase our control over reality. It enables us to translate the "external state of reality" into digital information, which allows us to use computers to edit, manipulate, share, and improve it, alter or update its context, and make it more valuable.

As mentioned earlier, the road along the path of Digitization began with numbers, then letters, advancing through imagery, audio, and video. In every case, their production, editability, distribution, and sharing became increasingly easier and more efficient and hence more valuable.

Spatialization is a technology that extends the extraordinary benefits and capabilities of digitization to every aspect of the physical world in which we live unlocking valuable new products, services and business models in the process. This is because spatial computing, like personal and mobile computing before it, has the rare ability to benefit all sectors of society—the consumer, public, private, and education sectors simultaneously.

Benefits extend to architects and industrial engineers who design, plan, visualize, and test their work, to scientists who simulate environments from the past, and city planners who need to simulate the impact of traffic in the future; in video games, TV, and movies to complex applications for healthcare and medicine, like simulations for training or assisting a surgeon in performing surgeries, to artists redefining our relationships to this new age.

Whether we are talking about shipping, trucking, and logistics companies that need to track shipments of assets geographically, or the ability of a mother to transfer a virtual character from the park across the street to her children's playroom, or the ability to rent a virtual Ferrari for a race across Monaco or the Moon, Spatial Computing humanizes our relationship with information.

With each generation of new computing technology, human to computer interfacing has continually evolved to be more and more natural and intuitive. Early interactions with computers required highly trained technicians; these days the average toddler has no problem interacting with the touchscreen of a smartphone or speaking directly to voice-controlled assistants like Alexa or Siri as if they were extended members of the family.

But what is driving this evolutionary trend and how does that impact the future of computing?

We could say that advancements in Human to Computer Interfacing (HCI) are ultimately biologically determined. From keyboards to mice, from alphanumeric displays to graphical user interfaces, and on to touchscreen displays of smartphones, we've seen a steady progression of our interfaces towards the more intuitive and natural. And now we are moving on to VR and AR or "spatial" interfaces as the next step, including voice, gaze, and gesture. Spatial Computing is not only a new technological advance—it is also deepening the link between the human brain and the computer brain.

We are not migrating to VR and AR merely because they are a fun, new technology, but because humans have binocular vision with depth perception, and these are the only interfaces that match our biology. They will increasingly become more useful, enabling us to perform more efficient and more effective interactions in the world, driven by the biology of the human brain and nervous system.

Our retinas contain an astounding 150 million light-sensitive rods and cones. Neurons dedicated to visual processing in our brains take up close to 30 percent of the cortex compared to 8 percent for touch and 3 percent for hearing.

But this is only part of the story. Humans respond to and process visual data better than any other type of data. Some statistics suggest that the brain processes images 60,000 times faster than text and that 90 percent of information processed by the brain is visual. What is true is that 30% of the brain is dedicated to the visual system. We recognize visual patterns very quickly and can respond much quicker than to words and numbers.

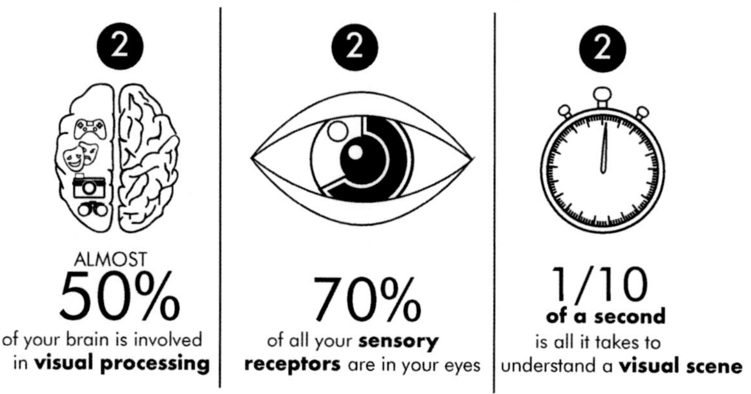

The recent multibillion-dollar investments in VR and AR technologies in order to advance towards an "eye-centric" interface tier is driven by the biological need for a 3D binocular interface. Anything else is just too inefficient. Ninety percent of the world's data was created in the last two years alone—and it is not slowing down.

The explosion of big data by a multitude of sources—from across websites and social media to the expanded use of mobile devices—makes it difficult for individuals and organizations to make sense of the data today, but as a new generation of wearables and IoT sensors are added, it will be nearly impossible to translate the oceans of real-time data into useful decision-making information, unless we upgrade the way this information is presented.

Zettabytes

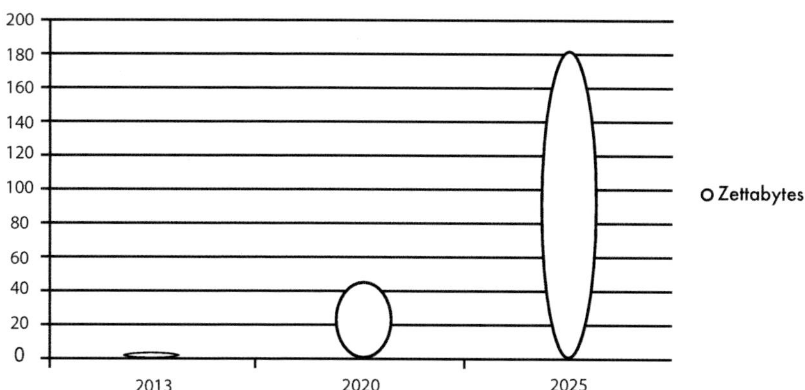

Spatial interfaces will be necessary in order to cope with this explosion of data. We need to be able to view it, navigate it, modify it, share it, make decisions about it, use it to simulate multiple alternative possible futures, and much more. The spreadsheet of 2025 will more likely be a simulation space that enables us to ask "What if?" questions and see the results displayed as a 3D immersive example of what we are testing or requesting. It's one example of the new model type of "the world," rather than that of "the book."

The language of humanity will also likely become more visual over time. The "memes" and emojis of Web 2.0 and the face masks and "lenses" of Web 2.5 companies like Snap and Tik Tok are heralds of this future.

The rise of Spatial Computing marks an essential next step in the evolution of our computing systems and highlights the importance of the Spatial Web in the ongoing evolution of computer-human interactions. During the years from 2020 to 2030, 5G mobile technologies will spread globally, giving us a global mobile network that can deliver low-latency spatial experiences. The spread of 5G network technology combined with the ever-decreasing cost and ever-increasing quality of spatial interface technologies will drive the global adoption of spatial not just because it is a more exciting interface—but because it is a biologically determined one.

As you can clearly see from these examples, the Spatial Web will make our world and how we interface with it, thousands of times more efficient than by using text and numbers. This will accelerate and improve education, create greater abundance in our economies and faster evolution of our technologies. History will see this as a State Change - similar to that moment when water turns to ice.

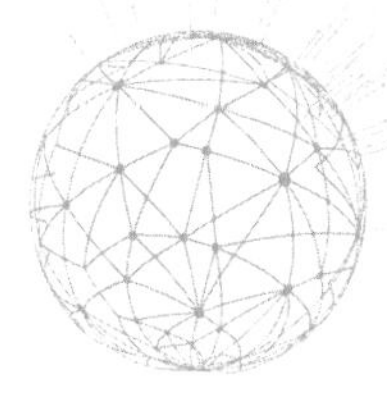

SPATIAL WEB TECHNOLOGY STACK

The prospect of a new, powerful, real-world web operating under the same technological frameworks and "surveillance capitalism" schemes as our current web is a recipe for disaster. Instead of our websites being hacked, it will be our homes, offices, drones, cars, robots, senses and biology that get hacked. The protocols at the heart of the current web, with their overall logic and architecture, were not designed for and not adequate for the emerging opportunities and risks that this new web of the world makes possible.

To enable the promise of the Spatial Web and address the shortcomings of the old web, a new set of spatial web protocols and standards for a new multi-dimensional web are required. We need a well-defined and robust specification for a comprehensive suite of new protocols and standards that support the trends of spatial, cognitive, physical, and distributed computing. We need a specification that is capable of laying the foundations for a web that natively supports universal values of privacy, security, trust, and interoperability by design, by default, from the foundation up—a specification designed to become a universal standard for people, things, and currency to seamlessly move between spaces, both real and virtual.

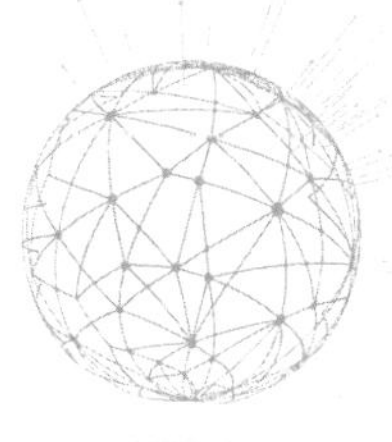

THE WEB 3.0 STACK OVERVIEW

The Web 3.0 era will not be defined by any single, individual technology, but rather by an integrated "stack" of computing technologies known in classic computer science as a three-tier architecture, comprised of Interface, Logic, and Data Tiers.

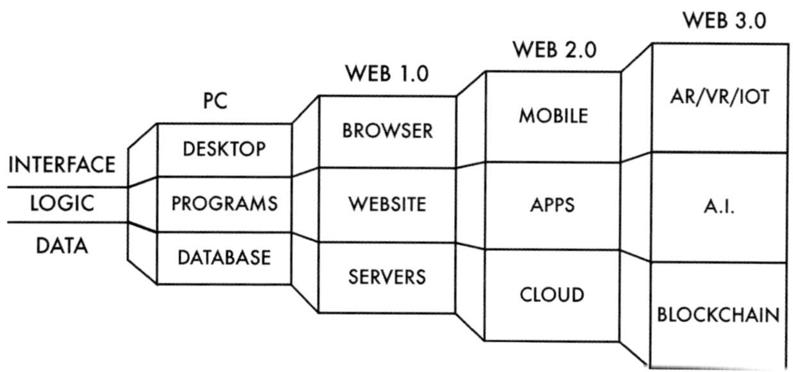

Web 3.0 will utilize Spatial (AR, VR, MR), Physical (IoT, Wearables, Robotics), Cognitive (ML, AI,) and Distributed (Blockchain, Edge) computing technologies, simultaneously as part of an integrated stack. These four computing trends make up the three tiers of Web 3.0.

Interface Tier: Spatial - Computing that takes place in a spatial environment, typically with special peripherals like AR or VR headsets, smart glasses, and haptic devices used to see, say, gesture and touch digital content and objects. Spatial Computing allows us to interface with computers naturally, in the most intuitive ways, best aligned with our biology and physiology.

Interface Tier: Physical - Computing embedded into objects, including sensors, wearables, robotics, and other IoT devices. This enables computers to see, hear, feel, smell and touch and move things in the world. Physical Computing will allow us to interface with computers everywhere in the world and receive information and even send "actions" into environments.

Logic Tier: Cognitive - Computing that models and mimics human thought processes, including smart contracts, machine, and deep learning, neural networks, AI and even Quantum computing. It enables the automation, simulation and optimization of activities, operations and processes, from production in factories to self-driving cars, while also augmenting and assisting in human decision making.

Data Tier: Distributed - Computing that is shared across and between many devices that each participate in a portion of the computer storage like blockchains and distributed ledgers or computer processing like edge and mesh computing. in general, this provides greater quality, speed, security and trust for the massive amounts of data storage and processing that are required for the Spatial Web.

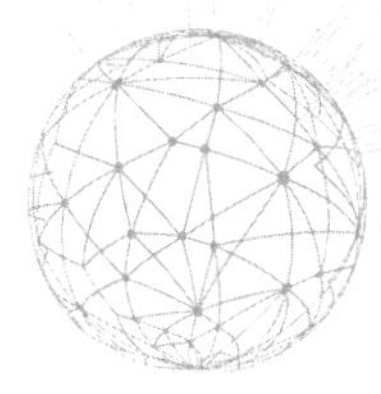

THE WEB 3.0 STACK IN DETAIL

The Interface Tier: Spatial Computing

Virtual, Augmented and Mixed Reality

Spatial Computing is a way of seeing and interacting with digital information, content, and objects in 3D space in the most physically natural and intuitive ways.

Every 15 years or so, a new computing interface emerges and dominates our interactions with computers: the desktop PC in the '80s, the web browser in the mid-'90s, touchscreen smartphones in the 2010s. Spatial computing technologies bring a fundamental change to the computer interface.

Three significant "Ages" define human interaction with information at scale: the First Age was the shift from spoken language to the invention of writing. The Second Age was triggered by the invention of the printed word (from written to printed). And the Third Age was the screen (from physical to digital). Each of these Ages radically shifted our economics, politics, and society. You may recognize these eras under the more familiar terms of the Agricultural, Industrial, and Information Ages, respectively. But viewing these Ages as evolutionary shifts brought about by advancements in our relationship with information highlights the importance of this next Age. Spatial technologies are the next evolution of the interface, progressively moving our attention away from the screen and into the world around us. This will have a far greater impact at a greater scale than any of the previous Ages.

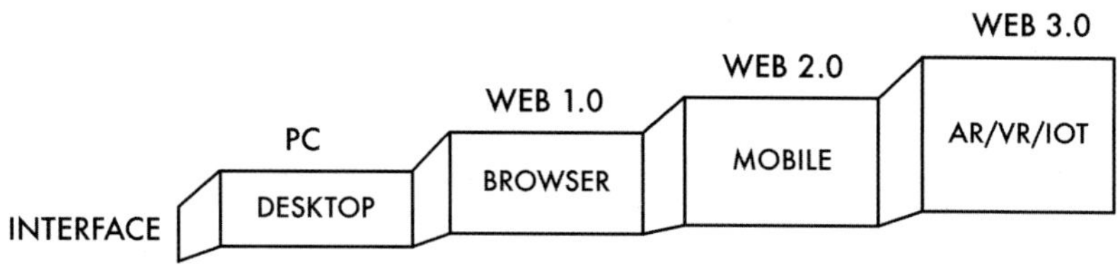

Our most direct experience of Web 3.0 will be via its interface. With Spatial Computing, the interface is literally the entire world, with data displayed everywhere, all around us, allowing us to interact with it via speech, thought, touch, and gesture, adding a new dimension to our information, ideas, and imaginations, enabling them to be immersive, collaborative.

First, let's look at the nuances of the main types of Spatial Computing.

Virtual Reality (VR) is a form of technology that allows a person to experience being somewhere else. It produces images, sounds, and even sensations to create an immersive sensory experience so that a user feels like they are really present in another place. That other place can be a virtual tour in another country, for instance, or a VR world like No Man's Sky or any combination of the real and virtual—sometimes called Mixed Reality (MR). Immersion in virtual reality gives a sense of being physically present in a non-physical world. VR is enabling us to enter fully immersive simulations for education, training, prototyping, and entertainment.

In VR anything you dream of can be experienced. Put on a headset and experience being transported to anywhere in the physical world, the universe, or any fictional universe, at any point in history—past, present, or future. Experience the widest scope of possible situations and scenarios. Be yourself, or be any character you wish—big or small, young or old, human or…other. Enter an artery to watch white blood cells fight off an invading virus, or travel through space and time at light speed to watch the universe being born. VR is programmable imagination. It is unlimited in its experiential applications.

On the more practical side, VR can enable us to collaboratively iterate on a city plan, a home design, or construction worksite to alter design and layout. Designers can simulate the ideal user experience long before the first shovel hits the ground and the build-out begins. While legacy technologies also allow us to prototype with immersive tech, VR provides a more direct experience by being able to walk, fly, and interact with simulations and prototypes. As a result, we're going to get better creations of our homes, offices, cities, and products.

Immersive media may also allow us to feel closer to each other and connect personally to global issues such as humanitarian crises. VR can enable a form of telepresence that evokes the kind of empathetic and emotional responses usually reserved for when we are physically present. It offers an experience that is simply impossible in other mediums, granting us the magical power to step into a 3D replica of a 1,500-year-old cave full of Buddhist art, to be transported into the shoes of a Syrian child living in a Jordan refugee camp, or to watch the Notre Dame tower burn down from across the bridge. We feel connected not because we are any more Buddhist or Syrian or Parisian, but because the medium reminds us that we all share the experience of being human.

Augmented Reality (AR) differs from VR in that it shows the physical location that a person is in, but allows digital imagery, information, and 3D objects to be overlayed and displayed in the physical world. Digital content or objects can be linked spatially to physical objects. You can, for example, attach a maintenance document to a piece of equipment or hide a character in the living room to be discovered. Objects in AR can react dynamically to an environment in all of the ways that we expect physical objects to do (e.g., texture, lighting, etc.).

With AR, you can simply hold up your smartphone or (soon) don a pair of smart glasses while on vacation at the Colosseum in Rome and see it as it looked in 200 CE or watch a historical gladiator battle from the stands. You can visit Times Square and see all the Instagram photos, Facebook posts, and Yelp reviews from your friends from the actual location they were posted, or view virtual art in a real gallery, or see the actual food pop up on your menu in place of mere words. AR lets you try on a new pair of glasses, shoes, or a watch simply by selecting your preference and pointing at the relevant part of your body. You can travel to a foreign country and see all of the signs in your native language or add a layer over the world that allows you to see every building and person as if they came from Westeros, *Star Wars*, or the Victorian era.

AR allows maintenance workers, whether biological, algorithmic, or robotic, to view the maintenance history of equipment at a factory, mining site, farm, or power grid by querying the equipment itself to request that any related documents, plans, diagrams, reports, or analytics about a thing appear in 2D or 3D on the device itself. A home appliance or new car could offer an interactive tutorial. Industrial equipment could display diagnostic or maintenance history. A grocery store or an entire mall could offer a 3D map and navigation, not just appearing as a map on the screen of your phone, but displaying in the air in front of you, routing you, a delivery service person, or a robot picker through the ideal route to complete tasks. And the products themselves display all of their relevant information and even supply chain history to verify their organic origin, fair use or sustainable practices.

For the Enterprise, AR can significantly increase productivity. As it evolves, AR will be able to provide immersive step-by-step instructions for technicians, leading to time-saving and cost reduction through improved performance. AR makes work more accurate and work environments safer through effective, engaging simulation and training. The precise visualization of internal components of machines and their parts facilitates a greater depth of knowledge and comprehension by providing rich simulation of different scenarios.

Interface Tier: Physical Computing:

The Internet of Things or IoT

Physical Computing is a way of sensing and controlling the physical world with computers. It enables us to understand our relationship to the digital world via our computers' relationship with the physical world. Physical Computing is the sensory and muscular hardware layer of the Spatial Web.

We've entered the fourth wave of the Industrial Era. The first was powered by steam, the second by electricity, the third by computing, and the fourth by integrated networks of sensors, beacons, actuators, robotics, and machine learning. These "cyber-physical" systems—a central feature of "Industry 4.0—will power the smart grids, virtual power plants, smart homes, intelligent transportation, and smart cities of tomorrow. The IoT allows objects to be sensed or controlled remotely using the existing Internet network infrastructure which creates new opportunities for more direct integration between the physical world and computer-based systems. This will result in improved efficiency, accuracy, and economic benefit.

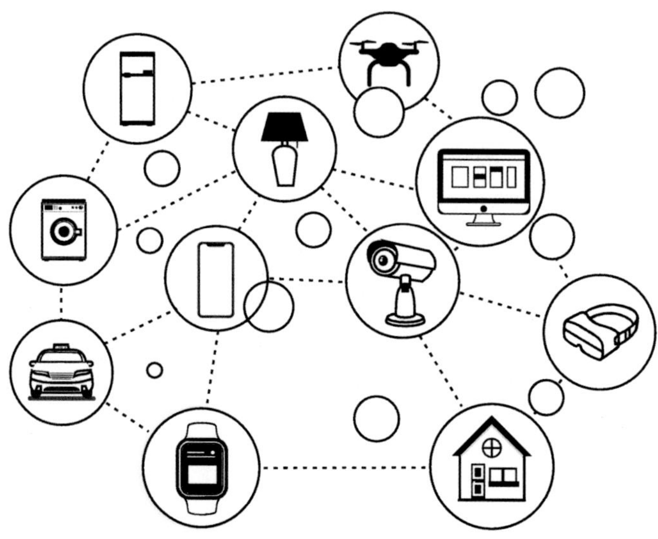

Just as we interface with the computer, in Web 3.0 the computer will interface with the world via the Internet of Things. "Things," in the IoT sense, can refer to a wide variety of devices including heart-monitoring implants, biochip transponders on farm animals, cameras streaming live feeds of wild animals in coastal waters, automobiles with built-in sensors, DNA analysis devices for environmental/food/pathogen monitoring, and field operation devices that assist firefighters in search and rescue operations. A more formal definition by Noto La Diego and Walden in their paper titled "Contracting for the 'Internet of Things': Looking into the Nest" describes the IoT as an "inextricable mixture of hardware, software, data and services." Generally speaking, it is a network of physical devices that are connected to the Internet and able to share data. These connected devices include sensors, smart materials, wearables, ingestibles, beacons, actuators, and robotics that will enable smart appliances, real-time health monitoring, autonomous vehicles, smart clothing, smart cities, and more to be interconnected, to exchange data, and perform activities in the world.

The Internet of Things will enable the digitization of every object in the world and capture data from every person, place, and thing. Think of this as the "read/write" Interface Tier to the planet. A trillion sensors will be laid over the world, like planetary-scale skin and senses with the ability to detect temperature, pressure, moisture, light, sound, motion, speed, position, chemicals, smoke, and more. This gives the IoT superhuman capabilities for good that allow these networked devices to see through walls to detect smoke in a highrise in New York, or sense the rising of a tide far in advance of a Tsunami in Indonesia, or the blood flow and pressure of an aging centenarian in Dubai, preventing the burning of a building, saving the citizens of an island paradise, and the life of a grandmother.

More Connected

Experts estimate that the IoT will consist of about 30 billion objects by 2020, growing to trillions of devices in the decades after. The evolutionary trend here is fundamentally more connected devices and more types of connected things. Effective application of this expanding capability can help us to use energy more efficiently, reduce carbon emissions, minimize waste, design better cities, predict diseases, track epidemics, and more.

The IoT as the Physical Computing hardware of the Spatial Web will capture and distribute physical data to the Cognitive Tier and store that Data at the Decentralized Data Tier via Blockchain, Edge, and Mesh networks for data storage and compute.

Logic Tier: Cognitive Computing

Artificial Intelligence, Smart Contracts, and Quantum Computing

Cognitive Computing is the digital application of the adaptive, contextual learning and logic systems modeled from our understanding of human cognition. These bring "smartness" into the physical world to analyze, optimize and prescribe activities in the Spatial Web.

The scale of intelligence:

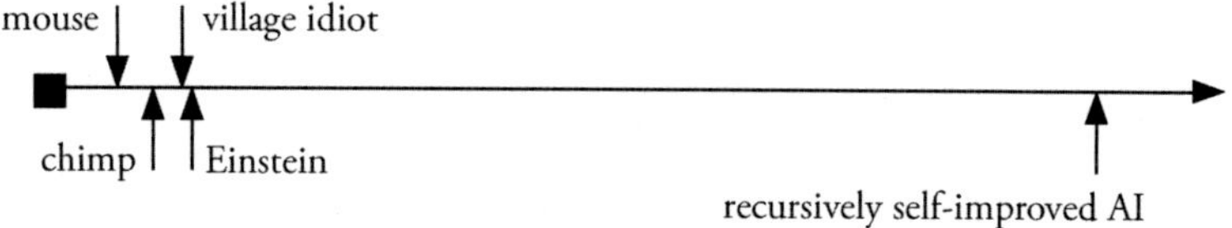

The Web 3.0 Logic Tier will be driven by the trend of Cognitive Computing in the form of several core technologies like Artificial Intelligence, Smart Contracts, and Quantum Computing. Populated by billions of self-executing smart contracts and programs, every building, room, object, and phenomenon will exhibit smart behaviors, and the environments around us will appear to have sentience. AI, Machine Learning, Smart Contracts, and related "cognitive" computing technologies shift us from the punch-card programs of early computers to autonomous, self-initiating, and self-learning agents that are adaptive. These will soon exceed the intelligence capabilities of humans but at the speed, scale, and scope of exponential technology.

Wikipedia defines the term "artificial intelligence" or AI as a machine that mimics human cognition. Many of the things we associate with human minds, such as learning and problem solving, can be done by computers. This is why everything in the future is frequently called "smart." It is a way of suggesting that it includes some programmable set of rules.

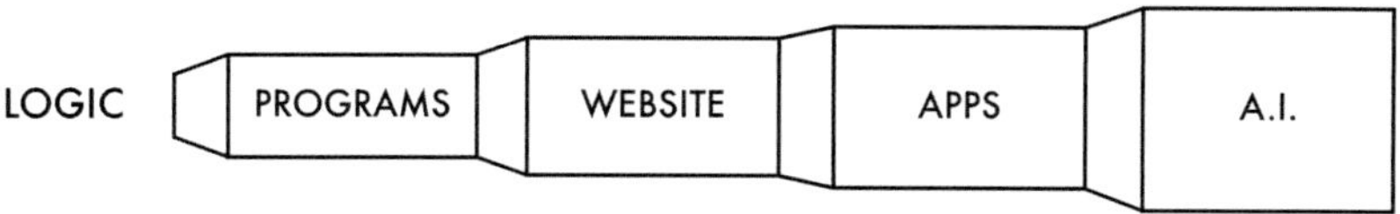

An ideal (perfect) AI is an autonomous dynamic agent that can perceive and act. It can see, hear, smell, touch, and even program its environment, modifying its behavior to maximize its chance of success at some defined goal. As a result, as AI becomes increasingly capable of mental, sensory, and physical abilities once thought to be exclusively human, our definition of "person" may need to be adapted.

Smart Contracts are "contracts as code"—they are programmable, automated, and self-executing software that removes legal contracts from the realm of documents that require constant human involvement and instead self-execute and self-enforce agreements between parties, provided the terms are met. If the program executing the contract is trustworthy, it's unnecessary to trust that the other party will fulfill the terms.

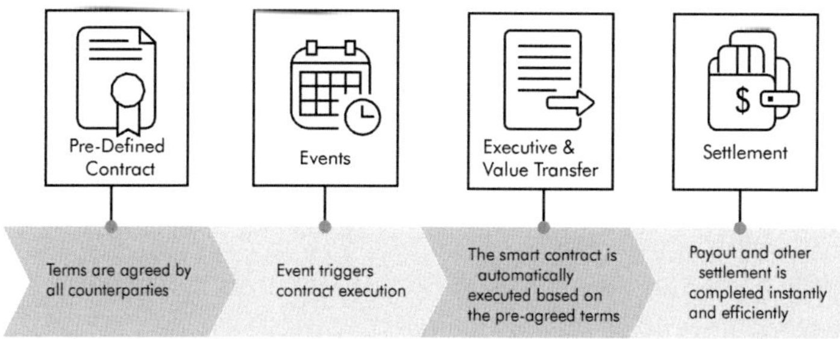

Due to the immutable nature of a Smart Contract's existence on Distributed Ledgers like blockchains, Smart Contracts provide security that is superior to traditional contract law and can reduce other transaction costs associated with contracting. Artificial Intelligence or AI will be capable of doing many things with Distributed Ledger Technologies, but for our purposes here, we will emphasize the role of AI as "smarter" contract agents that can enable data-driven, hyper-customized smart contract creation, analysis, execution, and enforcement.

Together, AI and Smart Contracts can simplify the negotiation and execution processes, while simultaneously facilitating more complex and dynamic agreements that can ultimately lead to greater efficiencies. This integral partnership launches the fields of law and software into an entirely new dimension. In the context of digital assets, smart contracts and AI can provide for the terms of use, payment, ownership transfer, and location-based terms or conditions automating entire supply chains including their transactions and segmentation of analytics for the data marketplaces of tomorrow.

Furthermore, Cognitive Computing will be applied to the vast amounts of data being pumped through the trillions of sensors of Web 3.0's IoT infrastructure. This will further augment and accelerate the human cognitive and creative processes across every domain and will increasingly allow AI to explore an unlimited number of possible futures.

Next, let's take a look at the role Quantum computing plays in the Logic Tier. Today's computers, called "classical" computers store information in binary, as either a "1" or a "0;" each "bit" is either on or off. Quantum computation uses quantum bits or *qubits*, which in addition to being possibly on or off, can be both on and off. Qubits can store a tremendous amount of information and utilize far less energy than a classical computer. By entering into this quantum area of computing where the traditional laws of physics no longer apply, we can compute significantly faster (a million or more times) than the classical computers that we use today.

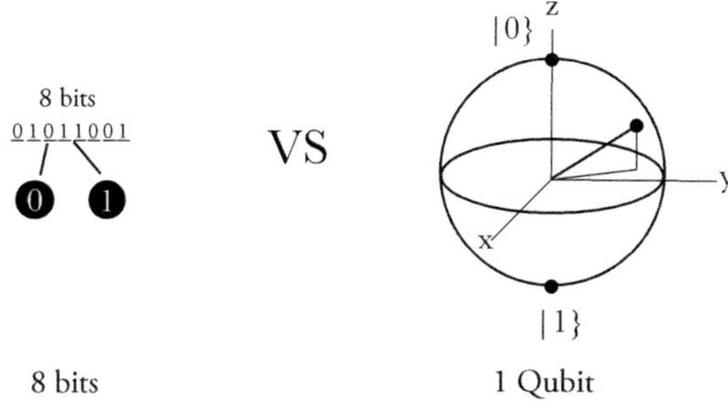

This gives Quantum Computers the ability to decipher the chaos patterns of traffic and the pulse of global markets, the nuances of the reflectivity of light as well as the neuronal activity of infants, the velocity of rain droplets, and the brush strokes of painters like a fortune teller reads tea leaves. It will seem so magical and impossible and yet it will likely uncover quantifiable patterns where we were sure none existed. Quantum Computing can act as a microscope for reality, revealing untold secrets of the universe capable of providing AI with the information necessary to make an infinite number of micro-adjustments to improve how our city traffic flows, or our children learn. It will help make VR even more realistic, route our resources where they are most needed, and might even inspire a level of appreciation for art in the most disinterested of Luddites.

The Cognitive Computing trend at the Logic Tier of the Spatial Web will make use of Distributed Ledger-secured Smart Contract logic and autonomous and adaptive AI as well as Quantum Computing, just as computer programs, web and mobile applications, and cloud computing drove the Logic Tier of the earlier iterations of the web.

Collectively, the power of Cognitive Computing technologies will intelligently automate every aspect of our personal and collective daily lives as well as operate our civil, social, political, and economic systems. In time, these algorithmic controls and micro-edits to our reality will appear to happen by themselves, occurring almost "automagically."

Greater Intelligence

In the beginning, we programmed computers in their language. Now they are speaking to us in our languages. They are seeing the world with their own eyes, and will soon apply Cognitive Computing to every aspect of our lives. Web 3.0 brings autonomous intelligence or "smartness" to everything.

Data Tier: Distributed Computing:

Distributed Ledgers and Edge Computing

Distributed Computing is the trend of pushing data storage and compute power closer and across multiple devices for speed and performance or farther away and more decentralized for greater trust.

At the bottom of the Web 3.0 Stack, at its Data Layer, are Distributed Ledger Technologies (DLT) like Blockchains and Directed Acyclic Graphs (DAGs)—decentralized and immutable ledgers—with the capacity to verify the provenance of information. DLTs like Blockchain offer a cryptographically secure, globally redundant method for storing records. These records are shared and updated across multiple computers (or nodes), distributed across the planet, and secured by cryptography. This creates a nearly unhackable, globally shared ledger of our records of events, activities, and transactions. Nodes can be financially incentivized to compete to validate each new record and punished if the data doesn't match against others across the network. With Blockchains, the most recent records and transactions are bundled into "blocks" of data and then added to the "chain" of previous blocks once their accuracy is validated by all the nodes in the network.

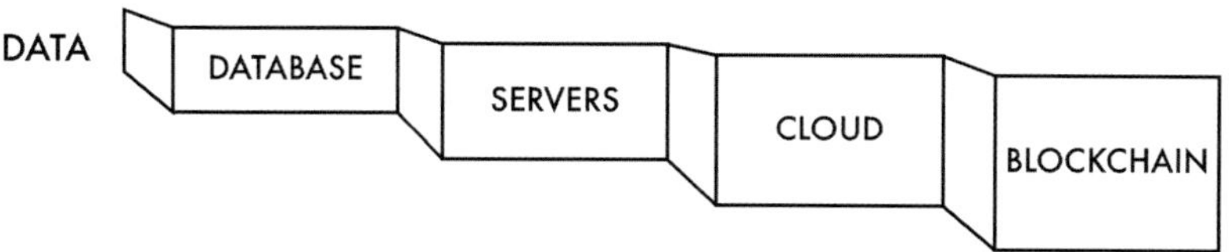

The Directed Acyclic Graph or DAG is another form of Distributed Ledger. A DAG is a network of individual transactions that are linked to multiple other transactions. DAGs trade the chain-of-blocks of transactions for a tree-like structure that uses branches to link one transaction to another, and to another, and so on. Some see DAGs as replacements for Blockchains, others as an enhancement. In either case, the combination of cryptography, social consensus, and innovative algorithms allow Distributed Ledger technologies to ensure "data provenance." A new generation of Blockchain startups has arisen to offer solutions to address the age-old problem of trust between humans. Today we see DLT-based solutions emerging for everything from global financial transactions to medical record storage, supply chain authentication to digital asset sales, and even shared custody of both physical and digital collectibles.

Distributed Ledger Technologies enable a world where every identity, contract, transaction, and currency can be trusted and verified. Trust emerges from the inherent architecture of Distributed Ledgers and does not need to rely on a corporation, government, or similar body to act as a trusted central authority. It promises a new economy and information marketplaces that could be genuinely open and decentralized. Like many new technologies, it is not immune from issues of standardization, scalability, and performance issues, but if history has proven anything, it is that if the need is great enough, then these problems are eventually addressed and overcome. And the need is great.

The Data Tier in Web 3.0 must be secured and trustworthy for the Spatial Web to work in the long run. Because the hyper-realistic, hyper-personalized, highly immersive and experiential "realities" that spatial technologies create (projecting our information and imaginations into the world itself, and displaying right before our eyes) also means that it will become increasingly difficult to accept the old adage that "seeing is believing."

Given recent advances in the computer-vision and rendering power of Artificial Intelligence and its ability to recognize and re-create everything from our faces, expressions, and voices to the objects and environments in the world around us, how we determine the *real* from the *unreal,* the true from the false, highlights the seriousness of Trust across the Spatial Web. Consider the fact that these technologies will not only have the ability to fake what we see or feel, the information and interactions of reality, but also will have the power to mine our information, influence us, advertise to us, and facilitate our transactions.

This poses a serious problem for our future, with individuals, societies, governments, and economies at risk. Critical data security foundations must be laid out in advance, followed by universal standards and policies that enable their adoption and support their enforcement. We must be able to reliably trust the who, what and "where" of our reality.

But how do we establish trust in a world where we can't trust our senses? Built on top of the current insecure web architecture, the potential for these technologies to be hijacked and abused by malevolent actors both human and algorithmic presents us with an unacceptable risk and a threat.

Since the dawn of civilization, humans have been trying to create reliable records about the things or assets they value. Our civilizations, economies, laws, and codes rely on records that we can trust. These records must provide reliable answers to the following critical questions about a valuable item such as:

What is it? Who owns it? What can be done with it? And...where is it?

Provenance is what makes a record "trustable." It is the historical record of the description, ownership, custody, and location of some-thing. Many of our technological and societal inventions like letters, numbers, bookkeeping, contracts, maps, laws, banks, and governments have emerged to address and manage the provenance of our records in the physical world.

However, empires fall, banks crash, and companies dissolve. Like them, our data records are all vulnerable to the passing of time, and like these institutions, many of our historical records have turned to dust. In the Information Age, we've progressed from paper records in cabinets to digital files stored in databases spread across the globe. In Web 2.0, more and more of the personal information collected online and via mobile applications has been stored on "cloud" databases by an ever-growing list of third-party companies that we have unwittingly given our trust to, and who are tracking and selling our data while leaving it vulnerable to hacks.

Now, with the arrival of Distributed Ledger Technologies as the Data Tier of Web 3.0, humans finally have a cryptographically secure, globally redundant method for storing and authenticating records. These records are shared and updated across multiple computers (or nodes), decentralized across the planet, and secured by cryptography.

This provides Data Provenance which enables unprecedented Data Integrity.

Edge Computing is another distributed computing paradigm. With Edge Computing, computation is largely or completely performed on distributed device nodes known as smart devices or edge devices as opposed to primarily taking place in a centralized cloud environment. The term "edge" refers to the geographic distribution of computing nodes in the network as Internet of Things devices, which are at the "edge" of an enterprise, a city, or other location. The idea is to provide server resources, data analysis, and compute resources closer to data collection sources and IoT systems such as smart sensors and actuators. Edge computing is seen as critical to the realization of physical computing, smart cities, ubiquitous computing, augmented reality and cloud gaming, and digital transactions.

More Trust and Access

From the siloed office database of the pre-web era to the globally accessible web servers of Web 1.0, to the mobile accessibility facilitated by the cloud infrastructure of Web 2.0, and on to the Distributed Ledger and Edge Computing that will secure AR information and power the IoT in Web 3.0, the evolutionary trend of the Data Tier is one of increased decentralization and democratization of data. At each stage, we have increased access and a "circle of trust" to include more and more participants at greater scales. This is the inherent value created by decentralized and distributed systems.

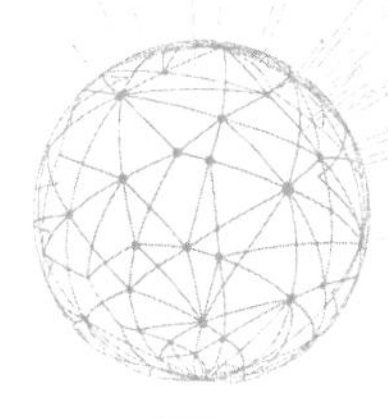

THE INTEGRATED WEB 3.0 STACK

Looking at the convergence of these various technologies through the lens of the Web 3.0 Stack makes it easier to see the benefits that can result from the *integration* of Spatial, Physical, Cognitive and Distributed technologies.

For example, in the Interface Tier, as the IoT provides us with sensor-enabled networks, Physical Computing will allow us to capture, measure, and communicate data regarding the performance of all physical activities. Robotics will perform any movement or transportation necessary in the physical world, from growing and picking our food to the manufacture and transportation of people and products globally.

Also in the Interface Tier, Spatial Computing as AR will provide the interface to a new world painted with a digital layer of information and contextual content that is constantly updated from the sensors of IoT, the intelligence of AI, and the new conditions set by smart contracts, secured by blockchains and incentivized by cryptocurrencies.

And Spatial Computing as VR will serve as a superior "pre-vis" experiential environment for the creation and exploration of our information, ideas, and imaginations. It will enable the most ideal virtual simulation or digital twin of any given object, environment, human or system.

In the Logic Tier, Cognitive Computing as Artificial Intelligence will provide analysis, prediction, and decision-making using Quantum Computing. Running simulations on our virtual digital twins will help determine ideal adaptations.

Also in the Logic Tier, Cognitive Computing as Smart Contracts can contextually govern, enforce, and execute all interactions and transactions via blockchain and distributed ledger networks informed the insights captured by the IoT and optimized by AI.

In the Data Tier, Distributed Computing as Distributed Ledgers and decentralized cryptocurrency platforms will maintain the trusted records for various people, places, things, and activities, and manage the storage and transfer of value across and between all parties. Distributed Computing as Edge and Mesh Networks will enable fast and powerful compute on location utilizing federated AI systems that can process information on device, sharing insights with the community while ensuring personal privacy.

The benefits of these various exponential technologies working together in an open and interoperable way is truly astounding. And it's because of this extraordinary potential that we, the authors, propose that Web 3.0 should be defined and described as a connected stack of technologies all working together as a part of a unified network—a network leading us across the various trends to The Spatial Web.

However, a network is not merely a converging stack of technologies. A key technology is necessary to enable the networks, something that connects them together and communicates between them. This technology is referred to as a "protocol" like Hypertext Transfer Protocol or HTTP that the world wide web uses to reference and communicate between the tiers of its stack.

The term protocol has many diverse definitions. In social etiquette, a protocol may refer to the acceptable behaviors in diplomacy or social affairs. A protocol in science is a predefined written procedural method of conducting experiments and in medicine it is the prescription that outlines how and when certain medicines or procedures should take place. In technology, a protocol is a common method for various objects or entities to communicate with each other. A cryptographic protocol is a method for encrypting messages. A blockchain protocol is a method for programming the consensus of datasets. A communication protocol is a defined set of rules and regulations that determine how data is transmitted in telecommunications and computer networking.

Each of these individual definitions of a protocol are designed to address various aspects of our biological, social, and technological lives. None are designed to replace the other nor interoperate or communicate with the others. This begs the question, what is the ideal intercommunication protocol to connect our biological, physical, and digital universes and turn the Convergence into a Network? We think the answer should be: a protocol that is designed specifically for the multi-dimensional needs of our future. Not the 30-year old protocols of the World Wide Web.

PROBLEMS

"Problems are only opportunities
in work clothes."

Henri Kaiser

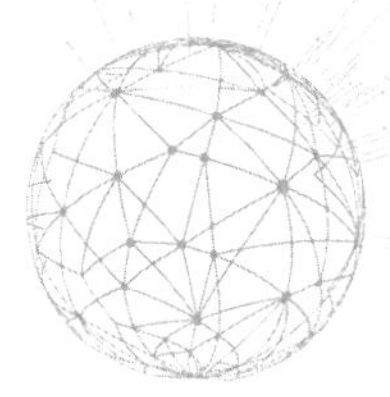

WORLD WIDE WEB LIMITATIONS

n 1997, in an article entitled "Realizing the Full Potential of the Web," the inventor of the World Wide Web, Tim Berners-Lee, shared his thoughts and hopes for the Web:

"The Web was designed to be a universal space of information, so when you make a bookmark or a hypertext link, you should be able to make that link to absolutely any piece of information that can be accessed using networks. The universality is essential to the Web: it loses its power if there are certain types of things to which you can't link."

This quote highlights both the larger potential and persistent limitations of the original Web protocols. The limitations are inherent in the definition of the Web itself as a universal "information space" designed for interacting with text and media on pages. Its greater potential is best reflected in an easily overlooked section of the quote where he says "universality is essential to the Web: it loses its power if there are certain types of things to which you can't link." Whether intentionally or not, this sentence suggests that for the Web to reach its true potential, it must enable anything to be linked to it, not merely text and media. But 20 years later, the existing Web protocols still do not easily enable certain types of "things" to be "linked."

As a testament to the original architecture and agnostic design of web addresses, web links or URLs can be assigned to pages of content as well as to "things." These things can be things we use daily like smart speakers, appliances, wearable devices, and connected vehicles as well as industrial devices or objects like those used for smart factories, and shipping and logistics (collectively referred to as the "Internet of Things" or IoT).

However, web domains and URLs have a very significant limitation. They do not provide for any reference to where things are located, spatially. The Web was designed to locate text or media on pages, not things or people in places; there is no way to search, access, or move people or things between various "spaces"—across the physical world, in games and apps, or across virtual worlds. This lack of a spatial domain or spatial address makes it difficult to manage how people, robots, or content operate within a space, in any authenticated or compatible way.

The HTTP web protocols were limited in their scope and ambition to connecting computers, documents, and media. Given the limitations of its era, the design did not include universal standards for user accounts, asset IDs, security, permissions, or transactions. Furthermore, they are file-based, not spatially-based.

Whereas HTML and other web programming languages let developers create page-based interaction and transaction rules, there is no spatial programming language to create the use policies and user permissions for interactions, transactions, and transportation of digital or digitized physical content or objects across real-world or virtual spaces.

The World Wide Web and its protocols are missing many things we now realize we need for connection in the 21st century. For example, the existing web protocols do not handle or validate data storage, consider location-based rights protection, asset ownership, or transaction authentication. There is no built-in reliable method for identifying and authenticating people, places, or things and the activities between them. Because the vast majority of our data is now owned and controlled by third parties, we are constantly faced with security threats. The incentive for companies to capture data and centralize it in silos in order to monetize it is clear, but the problem is that the bigger the silo, the bigger the honeypot is for malicious entities. Centralized systems by their very nature are susceptible to being hacked, corrupted, altered, acquired, bankrupted, or even destroyed.

The World Wide Web technologies and user interfaces were designed for interactions with 2D text (hypertext) and the navigation of pages for digital information, not digital experiences and actions. They weren't designed with Spatialization in mind and so are an insufficient foundation upon which to develop the next generation of Spatial Web applications that must enable interactions and transactions with 3D objects and user navigation within and across 3D spaces. We need a Spatial Web.

Let's take a closer look at the key features that are missing from the original web architecture and lacking from the current definitions of Web 3.0.

The Web today lacks a native identity or account infrastructure. This forces users to authenticate themselves with each and every service provider to access a provider's service. This requires users to have separate accounts for different interaction modes: browsing, communicating, sharing, buying. As a result, all of the value of the data associated with such an account is owned, controlled, and monetized by third parties. This is the case with nearly every service on the entire Web.

The Web today lacks an open spatial browser, built on a standard spatial protocol, that all users can access. There is no support for multi-user interoperable searchability, viewability, interaction, transaction, and transportation of users, assets, or currency within or across physical or virtual spaces.

The Web today provides no reliable real-time validation of users, assets, and spaces, or their identity, ownership, and permissions for various interactions and transactions. Because of this, the risk of virtual asset and environment alterations is significant. A hacker can edit AR content in the physical world or change the value of an item in a VR world for their own benefit. They can change all or a part of an augmented or virtual scene, delete or edit or change the accuracy of critical display information at a nuclear plant, deface or damage environments, move objects, take over automated drones, vehicles, or robots, or inject inappropriate or even psychologically harmful software or content. They can impersonate a person, their agents, or avatars; all of these instances can occur with both virtual and real-world objects, content, people, and locations.

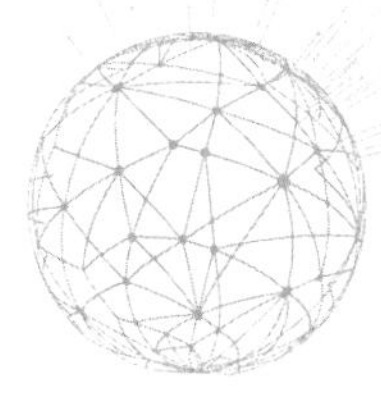

WEB 2.0 PROBLEMS

Hackers, Trackers, and Fakers—Today's Web is not secure.

Because the web was not designed with a shared database in mind, and because it has no login, it keeps no record, log, or "state" of our activity.

But as we browse the web, shop online, read posts, chat with friends, go for our morning jog, or park our car at night, our various digital and physical activity has been tracked and collected by various parties and this data has been stored on *their* servers. Weekly headlines across the planet continually report on how the largest and presumably most secure organizations in the world are regularly hacked, leading to identity and asset theft of an unprecedented scale. Even when not hacked, our personal data and activities have become a new kind of natural resource to be mined and sold to the highest bidder. Sadly, this has also led to the promotion of fake news and other counterfeit information posing as genuine articles. The monetization of our online and offline behavior actually promotes false stories over true ones because they fuel more user engagement. False rage "sells."

We have all been swayed by fake news in ways that have impacted us personally, professionally, and politically. As a global community, the full impact is yet to be truly understood. This is not merely a by-product of our gullibility or tribal affiliations, although both are arguably up for review. It is because the quantitative economic incentives to monetize the web by tracking our browsing activity, social relationships, and location history in order to better sell us products and services outweighs the qualitative moral implications to our society. The problem doesn't lie solely in the relentless pursuit of a dollar; but in the insidious nature of a kind of technologically-mediated monitoring and behavioral programming that lacks the appropriate level of care and concern for the psychological, political, or environmental cost-per-click. This is what happens when the invisible hand of the market is silently guided by an "all-seeing-eye" that remains blind to the effects its actions have on the world around it. The result? Surveillance capitalism.

But if you are a public company that must answer to your shareholders, whose only metric for success is your quarterly earnings or stock price, what do you do? If the DNA of your company is a by-product of a profit-by-any-means-necessary culture, how can you hope to change this?

Many people increasingly feel as if they are under a digitally-enhanced microscope that routinely tracks their most basic desires, and then targets these to make their fulfillment easier, more efficient and more effective—whether they want that or not. This is what happens when you apply Digitization to anything. You get exponential effects. Of course, this means that we can get exponentially good or exponentially bad effects. Which leads to another question: Are we applying Digitization correctly?

In an August 2018 article titled "The Man Who Created the World Wide Web," in *Vanity Fair*, the founder and inventor of the Web, Tim Berners-Lee, discussed with writer Katrina Booker the shortcomings of the Web's fundamental design and the crisis this produced in the Web 2.0 era.

"We demonstrated that the Web had failed instead of served humanity, as it was supposed to have done, and failed in many places." The increas-

ing centralization of the Web, he said, has "ended up producing—with no deliberate action of the people who designed the platform—a large-scale emergent phenomenon which is anti-human."

Maybe this is why it feels like the dystopian version of our future is imminent. We know deep down that something is very wrong and can see all of the signs playing out across our screens and slowly seeping into our daily life.

We can sense that a new era of the Web is approaching us like a hidden universe that will soon breach the veil of our current reality. And it is that feeling, the digital background buzz of anxiety, that leads us to a critical choice.

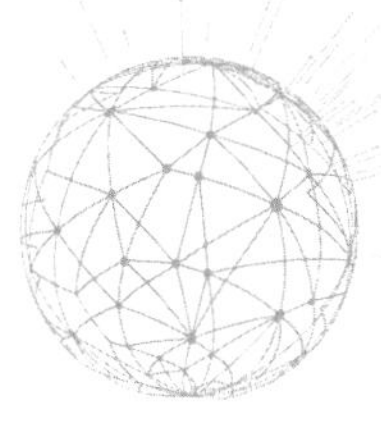

WEB 3.0 CRISIS/ OPPORTUNITY

n 2018, we reached 50% Internet connectivity on the planet. That is nearly four billion people all digitally connected and online, sharing everything from daily activities to political views to DNA information. Over the next decade, billions more will come online. But that's not all. In Web 3.0 *everything* comes online—trillions of objects—every appliance, every device or piece of equipment that runs or operates our farms and mines, our water and electrical plants, our cities and streets, our stores and homes, our forests and parks and our schools and government buildings will come online. Even our accessories and apparel, the watches and glasses and clothing we wear, are coming online. And billions of cameras positioned on every street light and building across our cities and on every drone in the air above us and every vehicle in the streets around us will also be in the hands or on the faces of our fellow citizens, every one of them with the ability to recognize who you are, what you are doing, how you are feeling, and maybe even what you are thinking.

In Web 2.0 we are often asked by developers to grant our mobile phones and apps the right to use anonymized diagnostic or location data in order to improve the performance of their app or services.

According to the Dec. 10, 2018 article, "Your Apps Know Where You Were Last Night, and They're Not Keeping It Secret" in the *New York Times*, "at least 75 companies receive anonymous, precise location data from apps whose users enable location services to get local news and weather or other information. Several of those businesses claim to track up to 200 million mobile devices in the United States—about half those in use last year." There are databases out there with close to a billion profiles. "The database reviewed by the *Times*—a sample of information gathered in 2017 and held by one company—reveals people's travels in startling detail, accurate to within a few yards and in some cases updated more than 14,000 times a day."

One dot shows a woman as she "leaves a house in upstate New York at 7 a.m. and travels to a middle school 14 miles away, staying until late afternoon each school day…The app tracked her as she went to a Weight Watchers meeting and to her dermatologist's office for a minor procedure. It followed her hiking with her dog and staying at her ex-boyfriend's house." Even though the records didn't disclose the woman's identity, the *Times* reporters easily connected her to the dot.

But what happens when many "dots" about you are being collected?

What level of information can be gleaned about you when someone is able to connect these dots as thousands of data companies and data brokers are doing across an entire shadow economy of Web 2.0?

In a 2019 *Washington Post* article titled "It's the middle of the night. Do you know who your iPhone is talking to?," the author Geoffrey Fowler reviews a privacy experiment that found 5,400 hidden app trackers operating on an average iPhone over the course of just one week. These various trackers shared personal details including address, name, email and cell carrier, location data as well as device name, model, ad identifier, memory size, and accelerometer data with third parties creating a treasure trove of personal data to be used for advertising, commercial, and political messaging. Most of this data was gathered at night when we are sleeping. Although some apps require tracking to be on in order to function properly, the experiment raises serious concerns about the transparent collection and use of consumer data. This is especially shocking given the strong stance and messaging regarding consumer data that Apple has taken.

The wearables, fitness apps, and connected home appliances of Web 3.0 will be reporting back far more information about us. Your devices will report details such as the number of cups of coffee you had at home, the content of your refrigerator, your mood before and after retrieving and putting back the ice cream container. Your toilet could tell the amount of fiber in your diet and so on. Your smart door could tell the exact time you leave and arrive home. While these various devices may collect genuinely relevant data for optimal performance of their particular app, or "app"liance, the collection and sale of that data and its potential enrichment when correlated with other datasets brings up unprecedented ethical and privacy concerns that industry and governments cannot ignore.

Wearable technology does offer amazing benefits and can revolutionize the healthcare industry. It enables us to use data to become more aware of our health in order to improve our well-being and prevent future health risks and illnesses. Fitness wearables address step tracking, sleep monitoring, and heart rate tracking, and even more complicated metrics such as diet, posture, skin temperature, and respiratory rate. They collect data regarding weight gain or loss, blood oxygen levels, and stress, which can be used to highlight potential risk factors or even alert others in real-time if life-threatening changes occur.

For both preventative and protective reasons, wearable technology will continue to be adopted. Apple has indicated that "Health" is their new frontier, and that is good because, where Apple goes, the rest of the market tends to follow.

However, the potential misuse of wearable health data for inappropriate monitoring, tracking, and classification increases the risk of discriminatory actions by insurance providers, employers, and governments and further allows these companies to exploit user data through third-party sales.

In a 2017 report in *Intersect* titled "Wary About Wearables: Potential for the Exploitation of Wearable Health Technology Through Employee Discrimination and Sales to Third Parties," the authors write: "There also remain ambiguous characterizations of wearable health devices as either electronic communication services or remote computing services and wearable health data as either content or non-content under the Stored Communications Act (SCA).

So is it content or speech or intellectual property or private property? At present, no one knows. The regulatory frameworks are far behind the market, chasing the problems that occurred half a decade ago or more. How will they protect us from the problems that will certainly arise in the decades to come?

All of this "lifestream" data can and will be mined for different outcomes. Together, we will feed trillions of bits of data into Web 3.0, making it more powerful, more valuable, and potentially more dangerous than ever.

Consider the near limitless analytical capabilities of AI applied to these kinds of data sets. With the ability to classify and comprehend the interactions between people, places, and things, it seems almost inconceivable but something like a "God View" of reality will be possible in Web 3.0. This could enable AI to make fairly accurate predictions about what might happen next in nearly any situation with a very high degree of certainty. As worrisome as this idea may be with its ethical and privacy implications and hints of *Minority Report*, this scenario would still be merely checkers to the chess game of reality when AI and

Quantum Computing reach their full maturity.

The converged power of AI and Quantum Computing would not only be capable of analyzing and making sense of any real-world scene but will be able to virtually *recreate* any scene imaginable and enable us to experience it. Don't like how yesterday turned out? You could re-experience it with a different outcome or even simulate a future tomorrow—in this world or any other where the people, places, things, and all the possible interactions would be computer-generated and could include simulations of the actual people in your life or imaginary ones. It would be like a choose-your-own-adventure for "reality" played out in a virtual reality that may be mediated by wearable glasses and haptic body suits or via a neural lace apparatus that connects directly to the brain. And all of this would be completely interactive and reactive in real-time.

The power of such technology for learning, experimentation, simulation, and more is incredible and has the ability to address some of our most pressing issues such as the climate crisis, poverty, inequality, and even racism by providing a level of visibility and detail on these issues that reveal new solutions. But there are real risks that are certain to emerge when we are not able to distinguish one "reality" from another any more than we can tell when we are in a very realistic dream. The human mind is incredibly susceptible to sensory suggestions and reality distortions. It happens to nearly all of us, every night. In Web 3.0, it will happen every day.

What is the cost of continuing down the path where our web and the world will become increasingly insecure and unsafe—where there is a virtual backdoor to every building, an open window to every room in our minds, where others can steal, alter, or delete our digital property including our identity, history, and means of communicating with others; where online identity theft grows up to become virtual impersonation, where psychological and biological hacking allows others to program our minds and bodies, our opinions, wants, and desires; where this kind of bio-social hacking actually becomes one of the largest and most profitable businesses in the future; where an elite few control the "master switch" to the web and therefore to our economy, society, our world, and our very reality? The cost, we know, is far too high.

If you think the current web has problems with hacking our data, tracking our actions, and giving us fake news, then the prospect of a powerful, real-world, geospatial Internet operating under the same ethical, technological, and economic design principles as our current web is a recipe for disaster. Instead of our websites and apps being hacked, it will be our homes, schools, drones, cars, robots, senses, biology, and brains being hacked.

This is just the beginning of the crisis/opportunity that we face in Web 3.0. Take avatar creation and authentication. In an era of massive ID theft, mega-security breaches, and private and public organizations failing to protect information, how do users authenticate the ownership of virtual avatars and digital assets? Which company can be trusted to generate and store them?

You wouldn't imagine exchanging personal information or buying goods online from an unknown entity today. In Web 3.0, the need for authenticated parties is even more critical. With virtual reality, the need to create virtual identities will become commonplace. In many cases, users will have different avatars for different uses, just as a LinkedIn profile and persona is usually far different from the same person's Facebook profile. Avatars will be some of the most important assets in Web 3.0 and for our daily lives. Their security and authentication will be of utmost importance to us. They will require a secure method to link back to us for biometric authentication.

Keep in mind that many of the avatars that will represent us won't appear as "us," although we may be acting and transacting through them. Some may be caricatures or cartoony versions of us, but others will be entirely different characters like the ones in games like Fortnite or virtual worlds like High Fidelity.

The launch of the iPhone X was a turning point for avatar technology. The next generation of smartphones and countless facial recognition cameras will come enhanced with 3D depth-scanning, emotion and expression recognition, and voice replication software that offer the feasibility of a hyper-real recreation of an avatar that looks, speaks, emotes and appears to act like you. What certification methods will ensure the authenticity and usage rights of our agents and avatars?

For a bit of context, a typical Web 2.0 identity problem would involve someone gaining access to the username and password of your Twitter account, but a Web 3.0 identity problem is when someone gains access to a complete, hyper-realistic replica of your face or voice or even biometric self. You can see the precursors of this type of digital impersonation with recent technologies like Deepfakes.

Deepfake and similar technologies use deep-learning AI to analyze photos and video of a particular person, say an actress or celebrity, and place their face onto another person's face within another scene. Quite often it is a woman whose likeness has been co-opted and placed over the face of a porn actress, entirely without their consent or ability to stop it from happening. And it's cheap enough for unscrupulous programmers to create in their spare time within hours. There are other less contemptible but equally creepy examples, such as placing Steve Buscemi's face on Jennifer Lawrence during an awards speech. While some of these are fascinating extensions of mash-up culture, watch Jordan Peele use his voice to drive a realistic video of Barack Obama calling for war with North Korea and that fascination will quickly evaporate, as the capacity to weaponize this technology becomes painfully clear.

For the time being, this is limited to video, but with volumetric video, generative AI, real-time 3D modeling, and increasing avatar use, it could be someone that is appearing to stand right in front of you, pretending to be someone you know, saying all the right things in just the same way as the person you trust would. But it's not them; the thing they are selling is fake, the environment isn't real, and they may not even be human, just a really smart malicious algorithm fed the right information about you to gain your trust to complete a transaction.

Given recent advancements in AI technologies, emotional detection, and gesture-recognition data combined with wearable sensor data, motion tracking, and other personal data including medical IOT information, it will become increasingly difficult to prevent the recreation of an indistinguishable replica of almost anyone, by almost anyone, for any purpose, within the next several years. How can we protect our identities, our assets, the content, and the spaces that we own or visit?

The main takeaway here is this. If the crisis of Web 2.0 was fake news, in Web 3.0 it will be fake reality.

If you're not scared yet, you should be. The collective power of the convergence of exponential technologies in the hands of humans, given our track record of abuses of power out of greed, malice, ignorance, and wishful thinking could lead to the end of life on this planet if not harnessed properly.

According to Wikipedia, the Doomsday Clock is a metaphor for threats to humanity from unchecked scientific and technical advances. It is a symbol that represents the likelihood of a human-made global catastrophe. Maintained since 1947 by the members of the Bulletin of the Atomic Scientists, the Clock represents the hypothetical global catastrophe as "midnight" and how close the world is to a global catastrophe as a number of "minutes" to midnight. As of 2019, the minute hand shows two minutes to midnight due to the twin threats of nuclear weapons and climate change, and the problem of those threats being "exacerbated this past year by the increased use of _information warfare_ to undermine democracy around the world, amplifying risk from these and other threats and putting the future of civilization in extraordinary danger."

The ominous tick...tick...tick...of the Doomsday Clock is wrong; it gives us a false sense of comfort. It gives us hope that there is still some time to "figure it out." But we are not two minutes away. The alarm bell is ringing already. The time to act is now.

In the Web 3.0 era, the crisis isn't limited to digitized information warfare. It includes digitized _experiential_ warfare that can quite literally impact us physically, psychologically, and even biologically because our information will no longer be trapped behind our screens, but will be placed into the world around us. It will be spatial.

SOLUTIONS

"The progressive development of mankind is vitally dependent on invention. It is the most important product of his creative brain."

Nikola Tesla

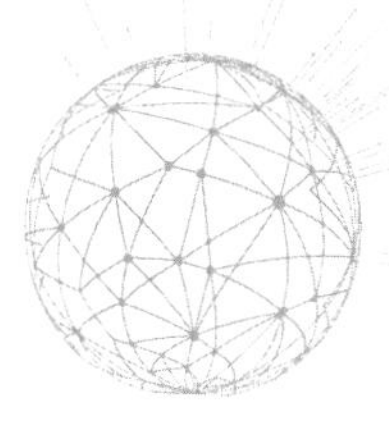

BUILDING THE
SPATIAL WEB

Today we need to move beyond the early web's ambition (and limitations) as a global network of interconnected computers, documents, and media. Web 3.0 is about building an intelligent and adaptive new web, a universal network of interconnected people, places, and things where we will be able to securely interact, transact, and share our ideas, information and imaginations. We need to build Web 3.0 to enable seamless communication for the transportation of goods and services from any point on the world to any other point in any world. In Web 3.0, we need to reliably trace origins from mine to market, from farm to table, and from game to virtual worlds. It must secure our virtual identities and their relevant profile information, activities, transactions, location histories, and digital inventories. Finally, it must enable a globally interoperable and interconnected digital economy that spans humans, machines, and virtual domains.

To fulfill this vision, the VERSES Foundation is proposing a set of universal standards and open protocols for Web 3.0 designed specifically to enable standards for defining and enforcing digital property ownership, data privacy and portability rights, user and location-based permissions, cross-device and content interoperability, and ecosystem marketplaces by enabling the registration and trustworthy authentication of users, digital and physical assets, and spaces using new standardized open formats, and shared asset indices secured by spatial domains, in which rights can be managed by a spatial programming language, viewed through spatial browsers, and connected via a spatial protocol.

In order to achieve a secure, unified digital and physical Spatial Web, we need a standardized way to identify People, Places, and Things (universal identity), a way to locate People, Places, and Things (universal address), a way to validate what we see and who we talk to (trusted data records) and an easy way to pay for goods and services anywhere we go across physical or virtual worlds (digital currency and web wallet). Most importantly, we need a way for all these things to seamlessly communicate together (a spatial programming language and protocol) that no one person, corporation, or government alone controls (open source) and a universal interface (spatial browser) that enables secure interoperable experiences across devices, operating systems, and locations across physical and virtual domains.

Why are universal standards important? They're important simply because we all want to work, play, and learn together better. We want to be able to interact, transact, and collaborate at home, at work, and in our schools and institutions across the globe. As we have seen with the original Web protocols, when we have a common technological vocabulary that we can use together, we can more easily and effectively communicate and share our information with each other and the world around us and, ultimately, those who will come after us. This is the benefit of networks.

The approach to this particular problem was more of a logical problem rather than a technological one, requiring us (the authors and developers) to consider our world systematically and holistically. As a logical problem, we needed to shift our way of thinking about both web technologies and the physical world around us as separate—and instead imagine them as intertwined or entangled. And we needed to consider requirements both multi-dimensionally and cross-dimensionally (from the real world to the virtual and back). This allowed us to evolve the existing concept of a web domain to that of a spatial domain, a web page to a web space, from a digital file, virtual object, or physical object to that of a smart asset and from a file-based protocol to one that is spatially-based.

This logical framework has allowed us to move beyond the early web's ambition (and limitations) as a global network of interconnected computers, 2D documents, and media to architect a more secure, intelligent, and adaptive new 3D web—a universal network of interconnected people, places, and things.

The Spatial Web requires a new set of technologies capable of enabling:

SPATIALITY: Digital content isn't only a single blob of data but is dimensionalized and inherently spatialized, making location the fundamental representation rather than strings.

OWNERSHIP: Users can own their data and digital property and choose with whom they share this data. Moreover, they retain control of it when they leave a given service provider.

SECURITY: Secure data collection, transmission, and storage enables interactions and transactions with virtual and physical assets between any user within and across any space—physical or virtual.

PRIVACY: Individual control, trust, and security utilizing cryptographically-secured and decentrally-stored digital identity enables "trustlessly" complete interactions and transactions with anonymity and auditability. Previously the exchange of personal data and layers of verification were required.

TRUST: Trust is based in reliable real-time validation of all users, assets, and spaces and their interactions with certifiable and verifiable records that validate various proofs of ownership, activity, traceability, and rights.

INTEROPERABILITY: Multi-user interoperability provides searchability, viewability, interaction, transaction, and transportation of any asset or user within or across any spaces. Seamless user navigation and asset transfer is enabled within and between spaces across devices, operating systems, and locations.

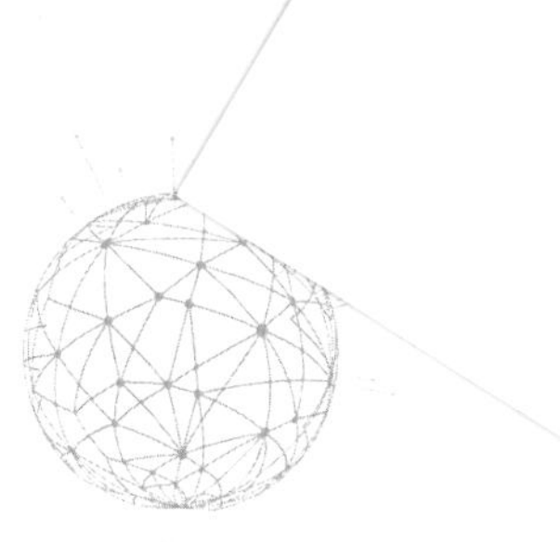

A SHARED REALITY

The Spatial Web requires new Spatial interfaces, Spatial Protocols and Spatial Programming Languages to interact with AR/VR, the IoT, AI, and Humans. Compatibility with Smart Contract and Distributed Ledger Technologies enables the validation of identity, ownership, and usage rights of any asset relative to its virtual and geospatial positions. This allows for the search, trade, transaction, trackability, and transfer of virtual assets by and between users within and across web spaces. The Spatial Web must be traversable and contiguous and be able to maintain geo and virtual position as well as asset and user-anchored persistence of spatial content.

The Spatial Web must

- Enable users to securely register, find, buy, sell, and transfer virtually anything between individuals within and across virtual web "spaces."

- Enable users to connect these spaces together to organically grow a Spatial Web that both visitors and virtual items can securely and reliably move between.

- Secure biometrically-authenticated human identities and virtual identities and their relevant profile information, transaction and location histories for representative agents and avatars.

- Enable location-based asset provenance, persistence and validation and allow assets to maintain and prove their uniqueness, ownership and history.

- Enable both a human and machine-readable understanding of the world and the ability for coordinated and collaborative biological, digital and virtual interaction.

But to actually enable this, we need that enables interoperability across all three tiers of the Web 3.0 Stack. We need a solution that allows for a "shared reality" at the Data, Logic, and Interface Tiers, simultaneously and in near real-time. A "shared reality" that is fundamentally spatial. One that is domain-specific, user-oriented and empirically discoverable whose data is reliable, verifiable, and secure. This is achieved and managed by a Shared Data Model maintained through all three tiers of the Web 3.0 Stack at the Interface Tier, the Logic Tier, and the Data Tier, uniquely connected by a Spatial Web Protocol that serves as the connective tissue and communication standard between all three tiers.

Interface Tier

Virtual, Augmented, and Mixed Reality "spatial" technologies collectively enable the Interface Tier to act as an Experiential Shared Data Layer. This occurs primarily at the Interface Tier of our Web 3.0 Stack and enables a shared experience layer. Readable and writable by humans, machines and AI.

Logic Tier

Spatial governance and business logic at the Shared Data Layer is created via a Spatial Programming Language that can be supplemented by any combination of smart contracts and AI, thus enabling a shared logic layer that can manage identification, permissions, credentials and validation for any interaction or transaction in any dimension.

Data Tier

The data processing and data storage layer writes, stores, and authenticates records across Local, Cloud, and Distributed Ledgers that enable a persistent, shared, and immutable data layer to allow for secure and reliable interactions and transactions.

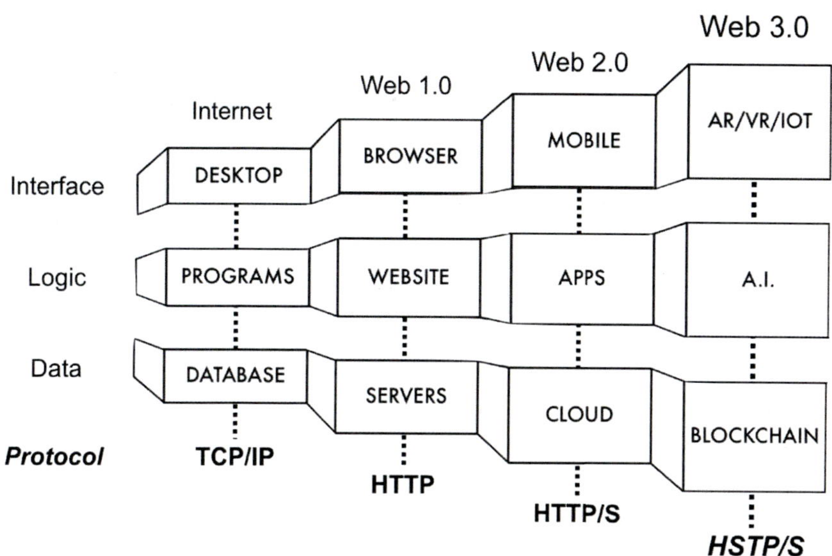

The Spatial Protocol

The HSTP protocol connects all three tiers of the Web 3.0 Stack. It is the key element in assuring that Web 3.0 is an open network and not a walled garden. It can reference real-world and virtual world coordinates, and securely record and authenticate the position of people, places, and things in a "cross-ledger" fashion. It references usage and ownership rights and enables virtual items to be both transactable and transferable in terms of ownership but also in terms of their relocatability within and across real-world and augmented spaces. It creates Smart Twins by enabling data to be linked and synced to physical objects, users and locations. The Spatial Protocol is what turns any space into a *web*space. It makes "space" smart.

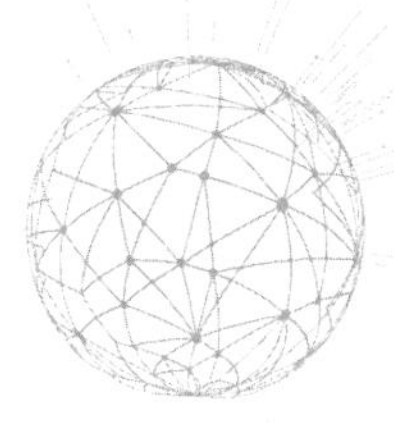

SPATIAL WEB STANDARDS

For Spatial For spatial content across both augmented and virtual reality to be searchable across any number of spaces—both geospatial and virtual worlds—and to enable multiple users to simultaneously view and interact with each other and assets across any number of device platforms —developers, creators, and users require universally standardized formats, languages, and protocol standards to facilitate these requirements.

Like the Web before it, the Spatial Web requires a standard approach to networking nodes that extend the concept of a node to be any physical or virtual thing in space that utilizes open standards for defining Identities, Addresses, Activities, and the ability to record and query for events or "States" that occur spatially. These are the key architectural requirements for the Spatial Web.

Domain - Address (Address and Ownership of a Space—Where)
Program (Rules for Who can do What, Where, When, and How)
Protocol (Communication between addresses—What to Where)
State (Records of Who did What, When, Where, and How)

Address	Program	Protocol	State
Web	HTML	HTTP	Stateless
Spatial	HSPL	HSTP	Stateful

Let's break them down so it is clear.

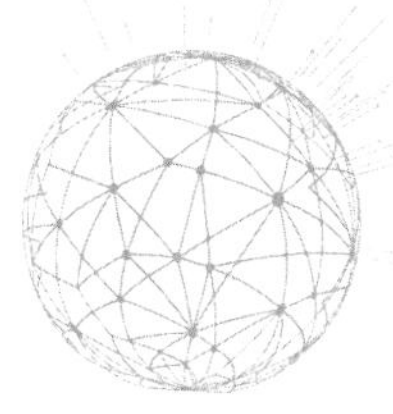

SPATIAL DOMAINS

A Web Domain name serves as a way to identify Internet resources, such as computers, networks, and services. Since an Internet Protocol or IP address for a computer is numerical in order for it to be machine readable, a text-based label was created to make it is easier for humans to memorize. A domain name is really just an address.

We have addresses to locate devices on the Internet, to navigate between pages on the web, or buildings via maps. At most, these require two dimensions. We exist and operate in the third dimension. We need digital addresses that can do the same. We need a solution that metaphorically AND literally "addresses" space. We need a three-dimensional or "spatial" domain.

The space around us doesn't have any universally acceptable and accessible "address." From the physical postal addresses that we use to correspond, navigate, ship, and receive mail and packages to the digital addresses that allow us to email, make phone calls and navigate between websites, none of them have a meaningful relationship to each other and present no method for integrating the physical and digital worlds. All current network-based address systems do not do a sufficient job at authenticating and networking people, places, and things.

Address examples:
Physical (postal) addresses for buildings
Telephone addresses (phone #) for our phones
Device (IP) addresses for our computers
Web (domains) addresses for web pages

In order to network spaces and then program the things and activities that occur within them, we first need a spatial address or spatial domain.

A Domain Name is an identifier in the form of an alpha-numerical phrase that represents an address. A Web Domain Name is a text-based name like amazon.com that points to an IP address such as 172.16.254.1 that is the address of the server that hosts a website. A Spatial Domain, however, points to a 3D volumetric spatial address made of coordinates that can be registered across Distributed Ledgers. Spatial Domain Names can be real places like "Joni's Café" or "Roman Colosseum," or virtual like "The Oasis" or "Hogwarts." Spatial Subdomains can be created to represent sub-spaces within the Spatial Domain. All domains confer spatial rights that can be digitally enforced. A Spatial Domain can control the permissions for AR content, IoT devices, Cameras, and Robots that are within its space.

Domain: A field, area, or range of action, influence, knowledge or responsibility.

Problem: There is no way to confer spatial rights, define permissions and activities of humans, machines, and AI within and across real and virtual locations because we do not have 3D volumetric addresses.

Solution: Enable Spatial Domains which are 3D spatial addresses that can reference any virtual environment or a physical location in the real world.

Benefit: Spatial Domains grant their owner the right to determine the rights, policies, and permissions of digital content and activity within their domain. These rights are much like the powers granted to web domain holders today with respect to the types of content access and user interactions that can occur on a webpage.

Example: Joni may define the coordinates of her café in New York as the Domain Space "Joni's," and then register individual rooms—kitchen, dining room, etc.—as subdomains. Joni can control the spatial permissions at her café for AR content, IoT devices, Cameras, and Robots.

In VR, the Oasis Spatial Domain can contain many subdomains within it such as the planets Archaide, Frobozz, and Ludus. Rights can be assigned accordingly.

Spatial Domain Registry

Similar to ICANN which manages the registration activities for web domain names, a Spatial Domain Registry will enable people and organizations to register Spatial Domain names and have them validated. For example, Paul and Norah may define their Home Domain as the coordinates and dimensions of their land in California, with the actual space of their home as Sub-Domains. Often the name of a space or business is used in multiple places around the world; for example, there's a Coliseum in Rome and Los Angeles. To keep Spatial Domains organized in a standardized way and allow for multiple instances of the name, the Spatial Domain registry will contain its address, including country, state, and city location and other profile information.

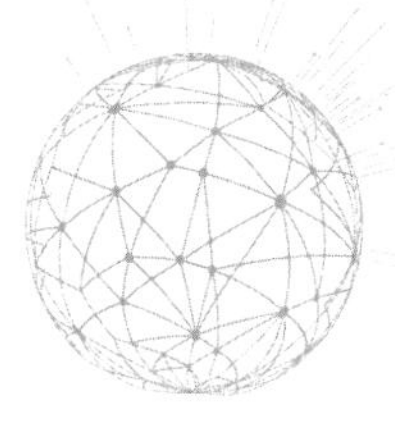

SPATIAL PROGRAMMING

HTML or HyperText Markup Language and later, Javascript, established standardized methods to "program" or layout and set the interaction rules for content on web pages. But with the birth of spatial computing and the need to present holographic content in three-dimensional space, none of the web-based markup or styling languages are sufficient tools to "program" space or validate spatial rules necessary for programmable interactions or transactions with assets.

HyperSpace Programming Language or
SPATIAL CONTRACTS

There are no standard rules for interactions or transactions between users, devices and locations with assets in spaces. Physical locations, objects and people have no standard method to describe spatial interaction rights, usage rules, searchability, or traceability of records. Therefore, you cannot search, discover, view, track, interact, transact and transfer assets between locations and users between virtual spaces and the real world in an open and standardized way.

HSPL or Spatial Contracts are "contracts as code"—programmable, automated and self-executing software that removes trade or service agreements from the realm of static documents that require constant human management and instead program them into the spatial interactions themselves so that contracts are executed by the correct action.

Solution: Enable a Spatial Programming Language or Spatial Contracts that can describe how to search, discover, view, track, interact, transact, and transfer assets between locations and users in virtual and real world spaces.

Benefit: Enables searching, viewing, tracking, interactions and transactions with assets and spaces, across time and space in both virtual and geo-locations. Enables objects to contain ownership, tracking, interaction, and transactional rules and records.

Example:

A contractor finds the 3D model of a building in AR and projects it into space where the building will be built to act as a real-world guide. Workers are guided through step by step spatial instructions. The completion of each task automatically fulfills a spatial contract. After the construction is completed, the foreman checks the model against the building structure and confirms it was built to specifications, down to the bolt, and the inspector confirms it was built to code. The contractor and associated parties are automatically and instantly paid upon approval of each step and phase.

While traveling in Tokyo, a father finds a rare Pokémon character and captures it for his young daughter. He teleports it to his home in LA for her to play with but limits its access to her bedroom.

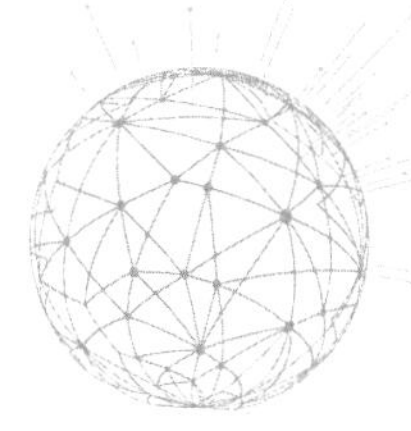

SPATIAL PROTOCOL

A UNIVERSAL SPATIAL PROTOCOL STANDARD

Web addresses designed for interactions with text, media, and the navigation of web pages are not a sufficient technological foundation with which to develop the next generation of spatial applications and web spaces for people and things to interact, transact with, and navigate between. It isn't designed to integrate the disparate technologies of the Web 3.0 era—AR, VR, AI, IoT, and DLTs. For this, we need to establish a new spatial protocol standard—one that communicates to and through each tier of the Web 3.0 stack. A spatial world needs a Spatial Protocol. It is the digital thread from which the Spatial Web is woven.

The Spatial Protocol can take into account the features, properties, and requirements of each tier of the stack, as well as effects generated by the entire stack working together. It is what enables Web 3.0 to be spatial, cognitive, physical, decentralized, and secure simultaneously, weaving each layer of the stack into a strong and contiguous digital fabric.

HyperSpace Transaction Protocol

The web is currently accessed by the protocol HTTP (Hyper-Text Transfer Protocol) to route users and content between web pages.

Similarly, HSTP (HyperSpace Transaction Protocol) will route users and 3D content such as objects/assets between web spaces.

The Spatial Protocol is not merely another disruptive technology but a foundational one, designed to serve as the digital infrastructure to support the next generation of Web 3.0 applications. Foundational technologies facilitate disruption across every industry. They enable users, objects (virtual or physical), and information to interact and transact together within and across any virtual or physical space, allowing these spaces to be networked together so that users, objects, and information can seamlessly move between them, like the Web. But unlike our current Web—a network of informational web pages locked behind a screen—a Spatial Protocol weaves a web for the world we live and operate in, a network of experiential web spaces—the Spatial Web.

HSTP is the solution for traversing the Spatial Web. Users and Smart Assets can be transferred or relocated between Spatial Domains anywhere across the Spatial Web, using HSTP for "hyperporting" between web spaces by allowing "hyperspace" links to be placed in one webspace that links to any information about anything or to another webspace. This is similar to how we link content and web pages on the Web today.

With an address for every space and a protocol to connect these spaces, we can move objects from space to space securely, we can track movements through space, and we can automate transactions through space.

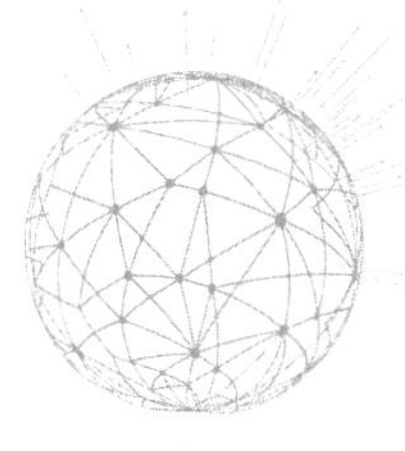

STATEFULNESS

Where the records are located is where the value accrues. The stateless nature of the World Wide Web lacked a method for secure, trusted, and shared data storage and access at the data tier. This allowed companies like Google to capture the "states" or series of events that users left as a function of their search index just as Amazon did with its sales index and Facebook with its social index, enabling them to monetize user actions, attention, and behavior. This caused trillions of dollars of value to consolidate around the application or logic tier and the service providers that enabled it. Users did not receive any of the commercial value they contributed to the network, nor were the developers of the web itself able to meaningfully monetize their creation.

The spatial index will be worth far more and should be part of a global public utility available to all. A "stateful" spatial index enabled by Distributed Ledger technologies will be critical to generating, representing, distributing, and securing value across The Spatial Web in a way not achieved on today's web. The tokenization of this protocol represents a historical opportunity to monetize the protocol of the new web that was absent when the World Wide Web was established.

A Spatial Index that is decentralized, censorship-resistant, immutable, and transparent allows users to retain the value they generate and the community to self-govern and monetize its creations while digital tokens can power and validate a vast range of spatial applications, interactions, and transactions. A Distributed Ledger-enabled and tokenized stateful web can reliably connect virtual and real-world spaces, track value, validate identity, and location while preserving privacy and data sovereignty. This allows for some powerful new features.

A stateful Spatial Web enables smart digital twins of people, physical spaces, and objects to be reliably and securely linked together, spatially. The effect of this is that when an object or person moves into or out of any physical or virtual space, a Spatial Contract can be executed automatically, subject to a set of spatial permissions set by the owner or approved entity triggering a record of the action and/or initiating a transaction. This makes the Spatial Web a trustworthy network for any form of interaction, transaction, or transportation. With Smart Spaces and Smart Assets, Artificial Intelligence will be able to see, hear, touch, and move things both physical and digital using computer vision, the IoT, and Robotics. It can be used to guide drones and automated vehicles, restrict access to users or robots, and track Smart Assets from one Smart Space to another. Smart Contracts and Smart Payments can enable seamless and secure interactions and transactions in both virtual and geo-locations, letting users with Smart Accounts transfer Smart Assets, pay for goods and services, and be paid for the use of yours, from world to world. This is the power of adding statefulness to the web. It enables entirely new categories of features and benefits. Including standardized identifiers for people, places and things allowing for hyperspace links to be added to the Smart Twin of any object or person to connect information about anything to the thing itself. Imagine a crowdsource-enabled "Wikipedia-like" dataset for anything, but linked to the object it is referencing in space not a page on the web. Turning any object into a kind of "Wiki-object."

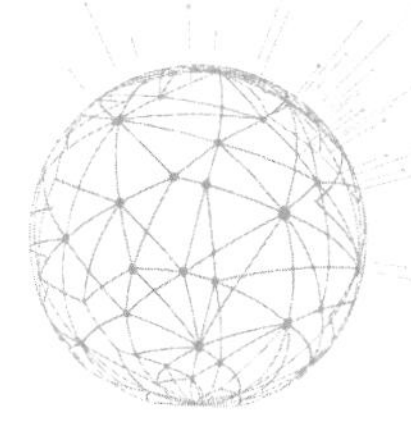

THE SPATIAL WEB COMPONENTS

The Spatial Web Protocol Suite is an open-source specification to enable a universal standard for users, assets, and currencies to seamlessly and interoperably move between virtual and geo-locations.

The Spatial Web Protocols and Standards necessary to enable an open, secure and interoperable Spatial Web should allow any real-world or virtual space to become a webspace, allowing users to track, interact and collaborate with spatial content or connected physical objects. The solution should enable interoperability across platforms, devices and locations and allow assets to be securely purchased and transferred between virtual and real-world web spaces, authenticated and validated, as needed by Distributed Ledger technologies.

The Protocol Suite can be described in five components: Smart Spaces, Smart Assets, Smart Contracts, Smart User Accounts, and the Spatial Protocol.

SMART SPACES

A Smart Space is a defined location— a virtual or physical "place" described by its boundaries, some descriptive and classification information and a Spatial Domain and set of interaction and transactional rules (Smart Contracts). Smart Space is "programmable space."

Smart Spaces are semantically aware and "know" via validation of transactions within them what Users or Assets are in them and can reference and validate the permissions related to those. Smart Spaces can be securely encrypted by Distributed Ledgers and can control what users, objects, software, or robotics are able to be used.

Problem: Currently, there is no way to reliably assign Spatial Rights or Permissions management for Users, AI, Spatial Content or IoT devices because there is no standard method to identify, locate, and assign permissions for activities spatially. There is no way to search for Spatial Content across real-world and virtual domains.

Solution: Enable any space to become a Smart Space whose boundaries are defined by coordinates—either real world (latitude, longitude, and elevation/altitude) with 0,0,0 or virtual (x/y/z) including outdoor and indoor spaces. Enable for sub-millimeter granularity and third-party re-localization optimization. Smart Spaces enable assets to have proof of their location, ownership, and permissions in time and space, across any device, platform, and location within virtual spaces and in the real world. They are searchable and can transact with Users or Assets. They can support multiple Users and Channels of Spatial Content.

Benefit: This solution enables multiple users to search, track, interact and collaborate with Smart Assets across time and space within Smart Spaces (i.e., in virtual and geo-locations). Smart Spaces are programmable.

Example:
A couple interested in buying a home in another state virtually

walks through the various rooms of potential houses and can place their furniture in it to see how it fits.

The Port of Long Beach notifies a buyer's account that the cargo that left Hong Kong has just arrived. The buyer's account automatically pays the shipper minus the port fees.

SMART ASSETS

A Smart Asset is any virtual or physical object that has proof of its unique existence, ownership, and location because it is registered on a Distributed Ledger with a single cryptographic ID containing its description, classifications, ownership, location, usage, and transaction terms, and unique history.

Problem: Objects have no universal provenance or record of history, ownership, location or permissions that are synced and linked to the object itself. They cannot be traded, shared, sold, or transferred between across locations, users, apps, games or virtual worlds.

Solution: Solution: Provide objects with universal provenance. Objects have proof of unique existence, ownership, and location are now Smart Assets.

Benefit: Smart Assets can be used, traded, shared, sold, and transferred subject to a set of programmable rules or Spatial Contracts across locations, users, apps, games, and virtual worlds

Example:

A woman searches across a thousand retailers from her living room and "tries on" a 3D virtual version of a watch, purse, and hat before placing her order.

A repair team follows an arrow in front of them across the roof of a casino to fix an air conditioning unit. Once there, they can see the maintenance history of that unit on the unit itself and be guided visually through the installation of a new part.

SMART (SPATIAL) CONTRACTS

Smart Contracts are "contracts as code"—programmable, automated and self-executing software that removes trade or service agreements from the realm of static documents that require constant human management.

Problem: There are no standard rules for interactions or transactions between users, devices, and locations with assets across locations. Physical spaces have no standard set of available rules for use. Therefore, you cannot search, discover, view, track, interact, transact and transfer assets between locations and users in virtual spaces and the real world.

Solution: Enable a set of programmable spatial rules or Spatial Contracts that determine who can search, discover, view, track, interact, transact, and transport assets between locations and transfer ownership between users in virtual spaces and the real world. Enable "connected" physical objects to contain ownership, tracking, interaction, and transactional rules and records.

Benefit: Enable searching, viewing, tracking, interactions, and transactions with Smart Assets and Smart Spaces, across time and space in virtual and geo-locations.

SMART PAYMENTS

Smart Payments enable programmable transactions that can authenticate and execute payments for any exchange of value between Users and Assets within and across Spaces in an automated way.

Problem: Because payments are not programmable, transactions require lengthy and expensive validation and settlement and cannot easily be automated in the physical world. This costs time and money. Also, a lack of micro-transactions restricts payments to high-cost items when certain interactions like those between machines may require small continuous payments i.e. "streamable money."

Solution: Use digital wallets and currencies for seamless, integrated automatable payments across any app, virtual space, or geo-location across Web 3.0. Enable micro-transactions for assets, experiences, and spaces and services. Allow Users, Assets, and Spaces to use wallets.

Benefit: Smart Payments enable automatable, seamless payments across the Spatial Web. Users, Spaces and Assets can transact with one another, subject to permissions. This allows Smart Assets to transact autonomously and enables micro-transactions globally at low or zero cost. This is not unlike the spatial transactions that occur when we exit an Uber car, enter an AirBnB home, our receive a Postmates order. The fundamental difference is that in the Spatial Web, these kinds of Spatial Transactions are a default benefit of its Smart Payment architecture. The outcome of such an architecture is a Spatial Network Economy that enables autonomic economic functions for any transaction.

Example:
A shopper walks into a store, is identified via facial recognition

cameras, grabs the items they desire and are charged to their Smart Account upon exit.

A gamer slays a famous dragon with a magic sword and then takes the sword into another world and sells to another player; the record of its dragon-slaying history goes with it. As the sword is used to slay various notable enemies, it is resold at an increasingly higher price across various game worlds.

SMART ACCOUNTS

A Smart Account is a single user account for use within the Spatial Browser and across any Web 3.0 application that authenticates, stores, and manages a User's Identities, Asset Inventory, User and Location history, Wallets, Currencies, and Payments.

Problem: Users have no single sign-in or account for the current web necessitating the use of multiple parties' accounts and sign-ins to create profiles to authenticate themselves in order to access third-party resources and services. Users do not own these accounts.

Solution: A single Smart Account for the Spatial Web that supports the W3C specification for Decentralized Identity (DIDs) enables Users to allow third parties to authenticate themselves to receive authorization to access User's Smart Account Profile information, Smart Assets and Smart Spaces, interaction, and transaction history based on User's privacy and permissions settings. Users own their own accounts, data, history, inventory, and content.

Benefit: Smart Accounts enable private, secure and seamless interactions and transactions of Assets, across Spaces and between Users.

Example:

A specific medical student in Cuba virtually attends an operation in Brussels led by a famous surgeon and then gets to perform the same surgery virtually, while being critiqued by the surgeon and peers from around the world.

An FAA-approved technician sees a 3D virtual copy or "digital twin" of a jet engine and its historical service record, makes repairs, and updates the service record.

FEATURES
&BENEFITS

"But what...is it good for?"

*Engineer at the Advanced Computing Systems Division
of IBM, 1968, commenting on the microchip*

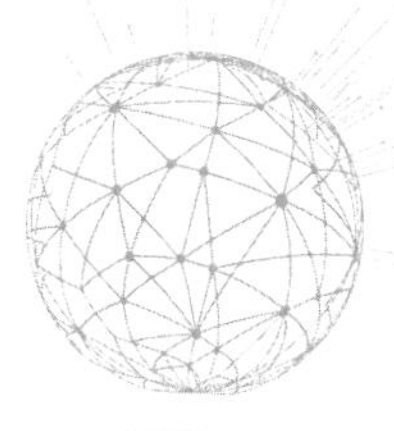

IDENTITY IN THE SPATIAL WEB

THE FUTURE OF IDENTITY

A verifiable and trusted identity is an essential necessity to enable the interactions, transactions, and transportation between people, places, or things.

Identification techniques have evolved over time, from beads and tattoos, written documents and printed passports, ID cards and birth certificates of yesterday to the cryptographic signatures and facial and iris biometric IDs of tomorrow. We use these tools in order to prove and assert who we are, where we come from, and what rights we carry with us. An identity is a social tool, one that fundamentally relies on trust in the system and method used to establish and verify the identity.

The White Paper entitled "On the Threshold of a Digital Identity Revolution" released by the World Economic Forum in January 2018 eloquently frames both the historical challenges and critical future needs.

"The issues associated with identity proofing—fraud, stolen credentials, and social exclusion—have challenged individuals throughout history. But, as the spheres in which we live and transact have grown, first geographically and now into the digital economy, the ways in which humans, devices and other entities interact are quickly evolving—and how we manage identity will have to change accordingly.

As we move into the Fourth Industrial Revolution and more transactions are conducted digitally, a digital representation of one's identity has become increasingly important; this applies to humans, devices, legal entities and beyond.

For humans, this proof of identity is a fundamental prerequisite to access critical services and participate in modern economic, social and political systems. For devices, their digital identity is critical in conducting transactions, especially as the devices will be able to transact relatively independent of humans in the near future.

For legal entities, the current state of identity management consists of inefficient manual processes that could benefit from new technologies and architecture to support digital growth. As the number of digital services, transactions and entities grow, it will be increasingly important to ensure the transactions take place in a secure and trusted network where each entity can be identified and authenticated.

Identity is the first step of every transaction between two or more parties. Over the ages, the majority of transactions between two identities has been mostly viewed in relation to the validation of a credential ('Is this genuine information?'), verification ('Does the information match the identity?') and authentication of identity ('Does this human/thing match the identity? Are you really who you claim to be?'). These questions have not changed over time, only the methods have changed." In the era of the Spatial Web, a historical opportunity has emerged that radically shifts the center of gravity concerning whom we are going to ask these questions to and who we will trust to answer them.

A UNIVERSAL SHIFT IN PERSPECTIVE

Occasionally, throughout history, revolutionary paradigm shifts occur that alter our understanding of the world and our place in it. In astronomy, in the mid-1500s a shift occurred that would spark the Scientific Revolution and set the stage for the Industrial and Information Ages to follow and literally redefine our place in the universe.

This shift was from the predominant view of the cosmos referred to today as the geocentric model which described a Universe with Earth at the center and the Sun, Moon, stars, and planets orbiting around it—to that of a heliocentric model in which the Earth and planets revolve around the Sun at the center of the Solar System. Although the idea that the Earth may rotate around the Sun, and not the other way round, had its roots in early Pythagorean thinking…it wasn't until Nicolaus Copernicus's *"On the Revolutions of the Heavenly Spheres"* in 1543 that the world became aware of this. Over the next century, the seed of heliocentrism grew its way through the Asian, Islamic, and European scientific communities until it finally blossomed with Galileo's proof, based on his observations of the cycles of the moon. For his historical discovery, Galileo was tried by the Inquisition, found guilty of heresy against the church and the biblical narrative of creation and humanity's place in it. He was forced to recant the facts of his discovery, on peril of death and was sentenced to house arrest for the remainder of his life.

What exactly was Galileo's crime? He had publicly placed his faith in objective and observable data and not in the subjective infallibility of the church. Galileo was one of the first to clearly state that the laws of nature are mathematical. He wrote "Philosophy is written in this grand book, the universe ... It is written in the language of mathematics, and its characters are triangles, circles, and other geometric figures." Galileo's groundbreaking scientific work and "trust in math" forged the path that led us to our greatest scientific inventions, eventually inspiring the work of Descartes and Thomas Newton whose achievements in philosophy and physics became the iron bedrock upon which the Industrial Revolution was ultimately built upon. Importantly, the key shift in the western world from the Scientific Revolution to the Industrial one was a shift from the faith in the centralized powers of the church to that of the decentralized trust in the verifiable power of science—in observable facts over unquestionable faith.

This change started with a willingness to shift perspective, coupled with an intense curiosity of the mind and an unquenchable desire to uncover a better way to describe reality more accurately and truthfully. Today, we have a similar opportunity to adjust our perspective in such a way as to create downstream changes that could alter our understanding of the world, ourselves, and each other for generations to come.

We will need to shift as we did before—from the geocentric notion that centrally-controlled 3rd parties are the ideal parties to validate, verify, and authenticate our identities and the data associated with them. We need to shift from a worldview in which we orbit around the respective planets of Web 1.0 and 2.0 service providers, subject to the gravity of their user agreements, ownership, and monetization schemes.

In this new worldview, our geo-centric perspective shifts to a heliocentric model where we discover ourselves to be like the sun, the source at the center of our own data system, where we verify, and authenticate ourselves, in a decentralized manner, where 3rd parties request access and permission to promote to us, to use our data, and where they orbit us.

Like the World Economic Forum white paper's title suggests, we are indeed on the threshold of a digital identity revolution. Much like the Scientific Revolution changed our model of the Universe from a geocentric to a heliocentric model, in the Digital Universe of Web 3.0

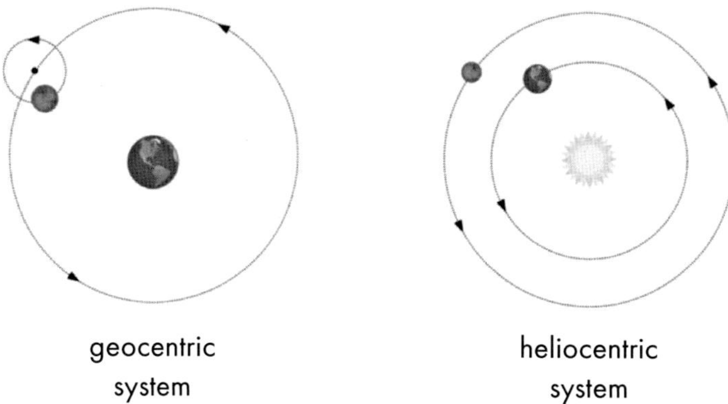

we must shift our center of gravity from a service-centric to a self-centric model for Digital Identity and all that comes with it.

Each person should be able to decide what information about themselves is collected as part of an online profile and of that information, they should have control over who has access to different aspects of it, and in what ways it can be used. Online identity should be maintained as a capability that gives the user many forms of control.

Blockchain technology offers the promise to revolutionize digital identity by returning ownership of personal data from companies and governments to individuals, such that individuals have the power to share their data with others and revoke it as they please as a human right.

TRUST IN MATH

Trust in the Webster-Merriam dictionary is described as "Assured reliance on the character, ability, strength, or truth of someone or something."

One can argue that Trust is the central orienting factor and organizing principle of our lives. Who or what we trust, when, how, and why determines much of what we call success in life. This is because Trust is the mediating force for any and all interactions. It is the scorecard of historical data combined with a map of trusted relationships and the proximity that a new one may have to a current or previous one. When we do not have direct historical data on a subject, we often rely on the proxies of relationships in individuals, organizations, and systems that we do trust. "Knownness" matters in the realm of Trust. The Mafia has understood this for years. It is often why family members are classically the most trusted and most likely to fail. We have a trust bias that lacks quantifiable validity.

Often based on proof of past performance, Trust's main purpose lies in its relevance to future activities. No one worries about trusting someone or something or some system in the past. You either trusted or you didn't, and determined trust was wise or it wasn't. In either case, the outcome has been written into the stone tablet of time and cannot be unwritten. Our concern lies in the future with respect to Trust.

For the better part of human history our fundamental trust, in reality, has been based on a combination of experience—in the form of empirical sensory data and in the ephemeral—and our interpretation of that data and how we attribute its causes. One might call it the Senses vs. Spirit debate. Millennia have been spent arguing about the veracity of the claims for the empirical vs. the ephemeral as the basis for trust in reality. But along the way, a third contender has quietly entered the field—digitization.

As our trust has increasingly shifted to digital formats, we don't so much trust our senses or our spirit guides as much as we trust the data.

Tesla drivers today have three options for driving.

1. Drive yourself. Trust your senses.
2. Let god drive. Trust the spirit.
3. Let the car drive. Trust the data.

Increasingly we will be relying on data to inform and drive the operations and activities of our world, our markets, our energy, transportation, health, and entertainment. Data integrity and our ability to interrogate its provenance and history will be central to our ability to trust anything of real importance in the future.

Finally, by allowing data ownership to move from companies and governments back to individuals, not only are individuals able to trust that the data they use can be validated through decentralized mechanisms but also organizations and governments can trust that they have reduced their own exposure to liability and risk associated with storage of such sensitive personal information.

PRIVACY BY DESIGN

The Spatial Web must enable "privacy by design." Privacy by design provides for individual control, trust, and security. And enables anonymity and auditability. It utilizes cryptographically-secured digital identity to "trustlessly" complete interactions and transactions that previously required the exchange of personal data and layers of verification. With the Spatial Web Protocol Suite the provenance of every thing, person, place, and transaction can be verified via an entry in a Distributed Ledger.

The Spatial Web identity architecture ensures that "privacy by design" is both a foundational principle and a core architectural imperative, where individuals possess the unalienable right to control their own digital identities. Each individual manages precisely what personal information is collected as part of an online profile or service, and defines who has access to this information, and the specific ways it may be used in any physical or virtual space.

Spatial Web account holders should have an absolute right to define bounded access to the complete contents of their own digital profile at any time. An online identity service can be maintained specifically to manage the many layers of control in a convenient and secure manner, as without a sufficient degree of flexibility and transparency, the trust in a system of federated network identity will be minimal.

INTEROPERABLE ID

A 21st-century digital identity system needs to create global identities, crossing international and virtual boundaries, without losing user control. Thanks to persistence and autonomy, global identities can then become continually available. Of course, these identifiers are not limited to humans. They can be applied to any and all people, places, and things, physical or virtual in nature.

To register the identity of any user, space, or asset on the Spatial Web, an individual or organization must first create an account and request a globally unique "decentralized identifier" (DID) based on W3C Standards. DIDs can be stored on a blockchain with quantum-resistant encrypted private keys which can be coupled with biometric markers and location-specific anchors to provide multi-factor spatial authentication, affording greater resiliency against Sybil and other types of attacks.

Think of them as super secure URLs for people, places and things instead of just for pages.

The number of IoT devices is estimated to reach 75 billion by 2025, and soon after will begin to approach the trillions. By standardizing data schemas facilitated by the trust enabled by DIDs, every IoT device can contain its own verifiable DID Activities of drones, cameras, cars or robots can be managed via spatial permissions. Essentially enabling a Spatial Contract to define that "these DIDs (drones) are allowed to be inside this DID (Spatial Domain of Santa Monica, CA) on this day/time/ weather etc." Having standardized IDs for any device in the IoT means that a global data commons and marketplace can arise. This will enable a global network of devices where spatially permissioned data can be transferred from machine to human and machine to machine. This free flow of securely-monetized data, functioning like a central nervous system for the planet, can enable ecosystems of efficiencies between humans, machines, and AI between communities, corporations, cities, and countries.

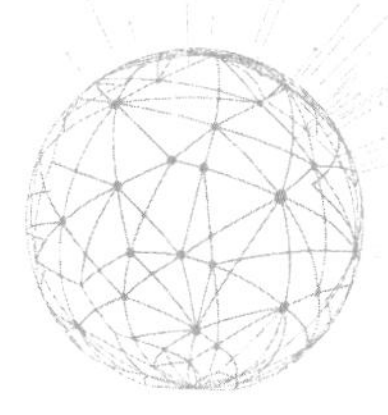

DIGITAL PROPERTY RIGHTS

Property can be physical or non-physical. By law, property is owned by a person, or jointly by a group of people or a legal entity like a corporation or even a society. Depending on the nature of the property, an owner of property has the legal right to consume, alter, share, redefine, rent, mortgage, sell, exchange, transfer, give away or destroy it, and to exclude others from doing these things.

Property rights were originally invented for land (real property) and then ideas (intellectual property). Until recently property rights did not extend to the digital world. The Spatial Web Protocol Suite provides a standard and open method that enables owners of digital property to assign property titles to various forms of digital data and then record the provenance of these titles on a blockchain. This creates secure digital property, and secondary markets around the digital property, which ultimately grows into a digital or virtual economy. Digital Rights are as critical to this emerging economy as physical and intellectual property rights were to our traditional agricultural and industrial economies.

The Spatial Web gives people the power to apply the same ownership principles that we have established in the physical world (e.g., owning land on which a house is built, owning a physical item to digital representations of themselves, their personal data, and 3D digital objects. Digital property, therefore, includes digital information related to our identities, avatars, virtual spaces, and digital assets including all of their interactions, transactions, contractual rights, and location history. When tokenized using a Distributed Ledger, physical or digital property can allow multiple parties to own fractions of the property or asset.

The Spatial Web blurs the lines between our well-understood concepts of physical and intellectual property rights, contractual rights, monetization, geo-fencing, and tradeability of digital goods in a way that requires us to reconsider their definitions and enforcement.

The Spatial Web enables a new economy for tradeable and transportable digitized assets that combine the virtual with the physical, allowing the digital world to leap off the screen and into the physical world with commercial models never before possible. With the integration of Distributed Ledger technology, a new generation of secure digital trade of assets and immersive experiences becomes possible.

Smart Property

As mentioned earlier, Smart Assets are Digital Property that use Distributed Ledgers to secure and control their issuance, ownership, and transfer independent of any centralized, third-party registry. Smart Assets can be anything—2D, 3D, digital, physical, virtual, people, animals, equipment, information, and more.

A Smart Asset is defined by a Universal Asset ID file that contains all the relevant information about that specific, unique asset—who, when, where, and how the asset can be used. And it is registered on a Distributed Ledger. A Smart Asset ID can refer to all of its relevant information and related files such as its 3D model ID and its metadata, including time of creation, value, descriptions, usage rules, etc. By registering this metadata on Distributed Ledgers, the Proof of Existence,

Authenticity of Ownership, and previous or current Geo or Virtual Locations can be determined and validated. This Smart Asset ID can also show relationship with other Assets (e.g., a Smart Asset could have smaller children and/or be part of a larger parent Smart Asset).

Smart Contracts can then be used to govern an asset's usage permissions, determining who can search, view, interact, transact, track, and transport it within or between Smart Spaces. Detailed information such as the relative coordinates, pose, and orientation someone may view a Smart Asset from can all be specified. Smart Assets contain reliable audit trails of ownership, location, and usage rules "within" the Asset itself.

"Real" Ownership of Digital Things

When we buy a digital product like a song, a video, or an app, the online store manages the database with our inventory. The product is licensed from the company and often limited in use to its proprietary platform of origin. The terms of their license and/or user agreement may mean that not only do they own the product, they also own all of our data associated with its use.

In comparison, if an individual purchases a Smart Asset, the individual owns it. Like physical products, the ownership is completely independent of the store it was bought in. A Smart Asset can be yours forever, and no one can take it away.

Digital Scarcity (The Double Spend Problem)

Collectibles, art, sculptures, coins, diamonds—people have long invested their money into scarce artifacts. Investment-grade assets in the physical world could be very expensive or impossible to replicate, hence the original could grow in value over time. However, it has traditionally been difficult to reliably prove a digital asset is scarce because it is just made up of computer code and can be copied endlessly at near-zero cost.

Computers gave us programmable, digital abundance, but we also need the ability to create programmable, digital scarcity, which enables us to imitate various real-world business models. Blockchains provide the solution by showing us exactly how many copies of an asset exist or could be issued at some later time. Blockchain-based property allows issuance rules to be transparent, removing the need to trust the issuer, and thus providing absolute confidence in the issuance data and the amount of Digital Scarcity.

Digital Provenance

The concept of Provenance originates in the fine art world where it describes the documented evidence that's used to prove that a work of art has not been altered, forged, reproduced, or stolen. Using the VERSES blockchain-based asset registries, the Spatial Web can now provide provenance by default to digital assets and spaces for the emerging digital economy.

Transferability and Transportability

Digital asset provenance enables the ownership of assets to be transferable between parties. Spatial Domain provenance allows assets to be transportable between locations. User Identity provenance enables users to transfer assets and themselves between real-world and virtual locations. This means that a hyperspace link can allow an object or user to "hyperport" to or from a virtual location like in the movie Ready Player One. In the physical world, Spatial Contracts can enable objects or users to be transported in the physical world much like Uber or Postmates automatically does today. Essentially having universal identifiers and addressability for any person, place, and thing allows you to transfer ownership and location.

Spatial Property Rights

One of the most important types of property rights in Web 3.0 will be the ownership and control of digital space. Whether this is the digital real estate of a physical property or a virtual one, similar rules must apply.

The freedom to buy, sell, and utilize property is protected in the United States by the 5th Amendment, where property ownership is viewed as a cornerstone of individual rights, economic growth and development, and the freedoms inherent to the society. Spatial Domains offer a form of digital Title that grants the holder absolute control over the digital use of their space and to whom, what, and when access can be granted or content displayed or sold. As more of the world is digitized, Spatial Property rights could become the most important property right in history, not only to control who or what can access our spaces but also what content can be displayed and even how and where transactions can be made.

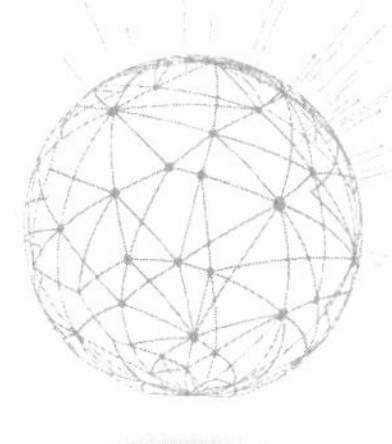

THE BIRTH OF DIGITAL COMMERCE

The early Internet's most valuable contribution was decentralizing *connectivity*—any computer could join the network by using standard Internet protocols, and thus, the Internet was born. Next came the creation of the World Wide Web in 1990 and a new set of "hypertext" protocols that further decentralized *communication* and led to an explosion of hypertext websites. In 2005-2006, Web 2.0—the Social, Mobile, Local web——enabled decentralized *content* creation, sharing and portability by putting it in the hands of users as a smartphone and distributing content through Facebook, Instagram, YouTube and other social media platforms. And with 2010 onward, we've seen the dawn of the next generation of decentralization, with the introduction of Distributed Ledger and cryptocurrency technologies.

Cryptocurrency

A cryptocurrency (or crypto currency) is a digital asset designed to work as a medium of exchange that uses strong cryptography to secure financial transactions, control the creation of additional units, and verify the transfer of assets. Cryptocurrencies use decentralized control as opposed to centralized digital currency and central banking systems.

The decentralized control of each cryptocurrency works through distributed ledger technology. Typically, a distributed ledger technology such as blockchain serves as a public financial transaction database.

Bitcoin, first released as open-source software in 2009, is generally considered the first decentralized cryptocurrency. Since the release of Bitcoin, thousands of altcoins (alternative variants of Bitcoin, or other cryptocurrencies) have been created. No one knows for certain if we will end up with one dominant cryptocurrency or millions of them for every conceivable category, converting back and forth in real-time across various global exchanges. But one thing is certain. A highly secure, digital medium of exchange that can be programmatically designed to suit the various transactions between human, machine, and virtual economies of the future means that the Internet has found its own kind of commerce.

Web 3.0 is poised to usher in the decentralization of trust, money, and the transferability of value itself (i.e., commerce). Until now, users had to leave the Web to complete a commerce transaction by going through centrally-controlled banking. With Web 3.0, *commerce* is finally a decentralized protocol and thus, digitally native.

	INTERNET	WEB 1.0	WEB 2.0	WEB 3.0
DECENTRALIZED				CONNECTIVITY
			CONNECTIVITY	COMMUNICATIONS
		CONNECTIVITY	COMMUNICATIONS	CONTENT
Online	CONNECTIVITY	COMMUNICATIONS	CONTENT	COMMERCE
CENTRALIZED	COMMUNICATIONS	CONTENT	COMMERCE	
	CONTENT	COMMERCE		
Offline	COMMERCE			

We've witnessed the decentralization of connectivity, communication, and content, but the Internet's latest trick is to convert commerce into a "native feature" of the Internet, woven into the virtual fabric of the Spatial Web with its own virtual money, exchangeable between any two wallets, by anyone or anything, and anywhere with no intermediary required. In Web 2.0 we have a global network where the Web interfaces with an external global economy. In Web 3.0 we get a distinct network economy, whereby the Web becomes its own economy.

Licklider's original vision of an "electronic commons open to all" is being realized in the Web 3.0 era by the Spatial Web—an open network that enables global access to the collective value (knowledge, power, wealth, etc.) of the world in its various forms. With connectivity, communication, content, and commerce, Web 3.0 provides the means to engage and exchange collective value between individuals anywhere on the planet. The Internet shifts value from centrally-controlled authorities to an authoring, distribution, and publishing network of peers. This is its core "value proposition." The Internet is a Decentralization Engine. But it is one that reaches its true potential in its next incarnation—Web 3.0.

In the 1990s, the Internet disrupted "communications" across industries and services like mail, publishing, telecommunications, travel, and even retail. It connected people, information, and businesses like never before. It was fundamentally about decentralizing the flow and access of information. The Internet presented itself as a kind of networked library of information accessed via a sort of digital book or "browser" with "web" pages.

In Web 2.0, the second wave of the Internet disrupted "content" like music, television, videos, and photos and enabled peer-to-peer sharing via social networking sites, blogs, wikis, video sharing platforms, and data storage sites. It also ushered in the Mobile Computing Era, which enabled location-based and crowd-sourced sharing technologies that disrupted "physical" services like transportation (Uber), accommodations (Airbnb), labor (TaskRabbit), and food delivery (Postmates).

The integration of geolocation (GPS) technologies into our smartphones, along with the introduction of mobile applications and operating systems, reshaped our digital lives. These new technologies gave us the ability to easily share and enjoy the content of the web, photos, and videos in near real-time from anywhere, and just like that, not only were our devices connected but we were connected. Perhaps more importantly, the elements of the "real-world"—people, places, and things—became digitally connected. "Local" searches blew past 60 percent of all search requests from smartphones.

Suddenly knowing where people, places, and things were meant that people could more easily get to places, and things could more easily get to people. This integration of the digital and physical happened so effortlessly, and yet has become so integral to our lives that we can't live without it. We also have yet to fully acknowledge this innovation. In this case, digital content was not the driver of these interactions; the relevance of content was now based on a far more critical factor that has emerged—context.

E-commerce powers online shopping. Over the last 20 years, e-commerce has grown to $3.5 trillion in annual spending. Some have argued that it is the fastest-growing economy in the history of the world. But e-commerce isn't really digital commerce and it isn't an economy. Why?

Because the authorization, storage, transmission and approval systems for payments are not "online." A long list of intermediary service providers, banks, gateways, and financial institutions utilizing 45-year-old telecom networks authorize and "route" your money and take their customary fees for this. International transfers can cost an additional 25 percent in fees alone. Besides, the currencies themselves are owned, controlled, and managed by states, central banks, and governments. Far too often, this can lead to currency manipulation, high inflation, onerous interest rates, devalued money, and bank bailouts. E-commerce is commerce where the transaction is initiated online, but where the ownership, storage, transmission, and records all occur off-line.

With Web 2.0 we have a global network that interfaces with an external global economy. E-commerce is a tourist that visits the web. In Web 3.0 we will have a network economy, whereby the economy itself is native to the Web—a Digital Economy.

We've witnessed the decentralization of computing, communication, and content, but the Internet's latest act is to convert commerce into a peer-to-peer exchangeable "native feature" of the web, woven right into the virtual fabric of the Spatial Web.

Janet Abbate, an Associate Professor of Science, Technology, and Society at Virginia Tech, in her seminal book *Inventing the Internet*, wrote, "People don't break into banks because they're not secure. They break into banks because that's where the money is." She went on to state that regarding the early designers and creators of the Internet, "They thought they were building a classroom, and it turned into a bank." But the Web wasn't designed to be a bank. E-commerce was a hack.

This hack has grown into a multi-trillion dollar value. Imagine the amount of value that we could create by intentionally supporting commerce at the core protocol layer of the Spatial Web.

Imagine all of the new ways that it might be used.

The Virtualization of Everything

In the Spatial Web, billions of new virtual assets, environments, and experiences will be created. In addition, trillions of digital sensors will be embedded into our appliances, cars, homes, and even our bodies. A massive number of new 3D-scanned "real-world" objects and locations will all become "virtualized." Characters, objects, and environments from history will leap from the pages of books and comics and the screens of television, movies, and games to surround us and walk among us. As a result, Virtual Assets will become the largest asset class in history. In order for us to benefit from their value, they will require a secure and interoperable means of proving their uniqueness and ownership as well as a means for inter-game and inter-world commerce and portability between locations—both virtual and real.

Digitally Designed 3D Models

Since the birth of 3D computer graphics in the early 1970s, billions of 3D models have been created. Programs from companies like Autodesk have dominated the creation tools for 3D models.

These tools and the models they create are used in nearly every aspect of our lives today across many different industries including television and motion pictures, video games, marketing, advertising, and virtual reality. They are also used for product design, for building design and architecture, civil engineering and city planning, and environmental and science simulations. If you look around the room you are currently in, chances are that the majority of objects around you were designed as 3D models on a computer. Turn on your TV and watch any advertisement or show or play any console videogame today and you will see computer-generated 3D objects, logos, environments, and characters everywhere you look. Everything from the Iron Man suit, to the Oral B toothbrush, to the latest iPhone, to next year's BMW is designed with computers; they are all 3D models.

Today there are numerous 3D asset stores with objects that number in the millions across platforms like Unity, TurboSquid, Sketchfab, and others. They all sell the pre-made objects, environments, characters, and more that populate our games, movies, and TV shows. But in Web 3.0, these assets can become transportable between the millions of AR apps and VR worlds that will make up the Spatial Web. Unlike much of our written words, photos, music, and movies, the billions of 3D models have one significant thing in common. They are not...online. Yet. They are all stored in siloed databases. They are not unique, easily sellable or distributed, and they are not yet a part of the Internet. But can you imagine the latent value that will be unleashed when they are?

Digitally Scanned and Twinned Models

The next category of Virtual Asset that has begun to emerge, and that threatens to eventually eclipse the Designed Models category, are the Scanned and Connected Virtual Assets. These are created through the use of scanning technologies that combine computer vision and depth-sensing cameras to create 3D models of pre-existing objects, environments, and even people. These 3D scanning technologies have historically been expensive to buy, complex to assemble, and cumbersome to use, but the latest smartphones now include these new cameras and AI chips in their hardware, and the software and features are built right into their OS. This will allow the next generation of users to create realistic scans of objects, people, and environments by default. Consider that with the next wave of smart glasses, drones, and automated vehicles with these real-time scanning technologies on board, you can see how many objects, environments, and people in the world will soon have their own hyper-real 3D models.

As more objects become computerized, we may see trillions of objects that are both hyper-real and connected. As mentioned earlier, the ownership and usage data for IoT and its connected devices should be secured on Distributed Ledgers—but how do we actually access that data and view and interact with these devices?

A Digital Twin is a 3D digital replica or representation of the data associated with a physical asset, process, or system. Arguably, any digital representation could be considered to be a part of a Digital Twin—from text-based diagnostic information to 2D blueprints, schematics, and pictures, to a complete 3D replica that represents all states, conditions, and history of any item (or even a human). However, most descriptions today trend toward a 3D or spatial representation. They are a part of the industrial and enterprise Digital Transformation evolution.

Think of a Digital Twin as a highly detailed virtual model that is the exact counterpart (or twin) of a physical thing. The "thing" could be a refrigerator, a vehicle, a human heart or even an entire system made up of a network of parts like a factory, retail store, or an entire city. Computer vision and connected sensors on the physical assets collect data that can be mapped onto the virtual model, allowing the Digital Twin to display critical information about how the physical thing is performing in the physical world, presenting current real-time state and activity as well as historical states.

Because a Digital Twin integrates historical data from past machine usage to display into its 3D digital model, a Digital Twin can contain an asset's entire history including origin, manufacturing, logistics, retail, home use, disposal, and repurposing. It can use sensor data to convey various aspects of its operating condition. And a Digital Twin can learn from other similar machines, from other similar fleets of machines, and from the larger systems and environment of which it may be a part. Digital twins can integrate artificial intelligence, machine learning, and software analytics with data to create living digital simulation models that update and change as their physical counterparts change.

Although the Digital Twin is predominantly used as a diagnostic tool to display a 3D view of the real-time information and historical lifecycle of an object, by adding AI, a Digital Twin can be used as a model on which to run simulations and predictive analysis. It can also serve as a holographic interface to the object itself, providing a human or AI a means to edit, update, or program its action. You can imagine a surgeon or technician using a Digital Twin of a robot to perform a remote surgery or to repair equipment.

In the Spatial Web, the fully realized expression of a Digital Twin is an IoT device, object, environment, or an avatar of a person that can be interfaced with a hologram via AR or VR, and manually and remotely controlled. If it is a physical object or machine, it can be automated via an AI whose actions are facilitated within Smart Spaces validated by Smart Contracts and actuated via Robotics. This can be referred to as a "Smart Twin" as all of its historical records and data are securely stored and reliably assessed via Distributed Ledgers, permissioned for various users, and monetized across data markets. <u>The Smart Twin is THE "killer app" of the Spatial Web because it uses all of the Web 3.0 stack of technologies.</u>

The implication of a Smart Twin for every person, place, and thing, every process, and state of all interactions, transactions, and movements across the planet integrated into a single interconnected network could lead to a 1:1 scale Smart Twin of the entire planet. A planetary-scale Smart Twin would be able to represent all of the uses of Earth's resources, the flows of all of its energies, all of the activities of its physical, economic, and social systems, all of the activities of its inhabitants—their hopes, dreams, attempts, failures, and successes. In the Web 3.0 era—"The World Becomes The Web."

But this is only one world and it's only in the physical domain. Even if we become a multi-planet species in the next century, that would only extend our Digital Twin to the physical galaxy. But Virtual Reality will already have created and filled entire universes from our imaginations.

World Builders and AI Generators

Minecraft is a popular sandbox video game that allows players to build things with a variety of digital blocks in a 3D procedurally-generated world, requiring design and participation from players. It is a kind of Lego-building play space on steroids. As of 2019, the important thing to note about Minecraft is that a young generation of 100 million kids has grown up designing and building an entire virtual world that collectively is nearly _eight times the size of planet Earth_.

This is not a mere children's game; it is a world-scale, civil engineering project disguised as a game. And it has created a generation of World-Builders.

Minecraft and other multiplayer worlds, e-sports, and games like the explosive Fortnite, with its dynamic avatars and fort-building community battles, have hundreds of millions of monthly users. They are generating billions in sales and they're inspiring an entire generation to build new worlds, objects, assets, and characters that have incredible utility and value to their communities. But these games, e-sports, virtual worlds, and MMORPGs are all siloed worlds. Users have no universal method to move between them or transfer objects and assets between them. But as the Spatial Web evolves it will be capable of establishing portability and rendering standards for objects, shaders, characters, and powers in such a way that the 2 billion global gamers will be able to find novel ways of linking, porting, mashing up, and building new worlds that work together.

But even billions of gamers and builders all working together to build millions of virtual worlds—Smart Spaces all interconnected as web spaces across the Spatial Web—will pale in comparison to what AI will soon be able to create.

Generative Design

Although computer-generated procedural graphics have been around for decades, the press really began taking notice with the release of No

Man's Sky, an action-adventure survival game released worldwide for PlayStation 4 and Microsoft Windows in 2016. The game uses an algorithmic procedural generation system to give the planets in the universe ecosystems, each with their own lifeforms and alien species for players to engage in combat or trade. How many planets does this game have? Over 18 quintillion unique planets. According to the game's Wikipedia page, "within a day of the game's official launch, …more than 10 million distinct species were registered by players, exceeding the estimated 8.7 million species identified to date on Earth."

Whether No Man's Sky or any other procedurally-generated game is able to attract and maintain players is anyone's guess. The thing to note here is the unbelievable scale of the power of procedural and generative algorithms to create immersive and dynamic experiences—literally out of thin air. But what happens when you combine procedural algorithmic technologies with AI?

Generative Adversarial Networks and Generative AI

Generative adversarial networks (GANs) are a class of artificial intelligence algorithms used in unsupervised machine learning, performed by two neural networks competing with each other. One network generates examples (generative) and the other evaluates them (discriminative). For example, a GAN can learn to generate pictures of trees. The generator attempts to create a tree. The discriminator has been given thousands of tree pictures to look at and knows what a tree should look like. It "fails" the generator until it actually learns "tree-ness" and can generate images of trees that can fool it. Over thousands or millions of attempts, the two refine the model until a result is produced that is remarkably accurate to reality.

GANs have been used to produce samples of photorealistic images for the purposes of visualizing new furniture, shoes, bags, and apparel items. And GANs used to modify video to "face-swap" one person for another in a scene are behind the Deepfakes category. This category of AI can be referred to as "generative" for its ability to reverse engineer a dataset, recognize and extract any pattern, style, form, or function that the data contains, and then generate an output.

Generative AI or GAI can mix, match, modify, edit, and algorithmically generate any image, sound, object, or scene, digitally rendering it as text, audio, video, or graphics, in 2D or 3D. It can even code it into software or 3D print it. GAIs can even be used to produce code to generate software for specific applications. Utilizing 3D printing, CRISPR gene-editing and related technologies, generative materials can render organic molecules, prosthetic limbs, and other items from scratch.

In the Spatial Web, Generative AI's ability to compose dynamic music, lighting, sound effects, conversations, and complex scenes will soon be used to auto-generate contextually meaningful narrative arcs, fully interactive and immersive AR and VR environments, architectures, products, and characters. By using many of the mood-tracking metrics and bio-markers available from our interfaces, these experiential environments can be completely personalized or made adaptive to social or environmental conditions.

The Spatial Web will change how we create art and culture, design and create products, build environments, enhance our bodies, and experience and share the realms of the imaginary. A new generation that opts to create, play, and work in a virtual universe will challenge our perception of place, economics, community, and self-worth. And AI's staggering ability to generate entire universes filled with unique environments, populated by intelligent characters and scenes that enable novel experiences at near-infinite scale will redefine the word "reality" for future generations.

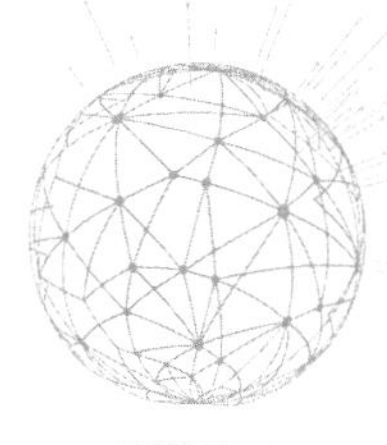

CONTEXTUAL IMMERSIVE ADVERTISING

THE DEATH SPIRAL OF ONLINE ADVERTISING

I t has been suggested that advertising is the thing that "ruined" the web. Many of the founders of the largest tech companies famously abhorred the concept of monetizing their services and applications with advertising; they wanted the functionality of the original service to drive engagement. Google, Facebook, YouTube, Instagram, Twitter, Snapchat, and WhatsApp all launched and drove massive user growth based on their core service. But aside from charging for their service, which would dramatically reduce their user base, selling advertising was the only other option to succeed financially online. Over time, the user data market emerged which gave advertisers the ability to more effectively target users. Ads became more and more targeted. As mentioned previously, Google built a Search Graph in Web 1.0, Facebook a Social Graph in Web 2.0, and they were able to monetize these in unprecedented ways.

Senator Orrin Hatch at a Congressional Hearing to address the Cambridge Analytica scandal and the potential manipulation of Facebook's ad platform by Russian spies asked "If a version of Facebook will always be free, how do you sustain a business model in which users don't pay for your service?" Mark Zuckerburg famously replied, "Senator, we run ads." History may look back and decide that this was the mantra of the Web 2.0 era.

The Threat of Hyper-Reality

The scale of the hyper-targeted personal advertising market in the Spatial Web is exponentially larger than in its predecessors but carries the same double-edged sword with economic value on one side and human values on the other.

Hyper-Reality is a concept film by Keiichi Matsuda that shows a day in the life of a woman immersed in a futuristic world where her vision is filled to overflowing with games, Internet services like Google, and various other functions, alongside an assault of advertisements that constantly pop up as she moves around the city. It is an artistic exercise in digital consumption; it is death by a thousand notifications. And it perfectly showcases everything that we do not want Web 3.0 to be.

Similarly, there is a scene in Steven Spielberg's film adaptation of the novel Ready Player One where the film's corporate antagonist shares their monetization strategy for the Oasis, an interconnected VR universe. "Our studies show that we can fill up to 80 percent of someone's visual field (with ads) before we induce a seizure," he says. Two better examples of the insane drive to monetize our field of view could not be made. What's of even greater concern is that breakthroughs in eye-tracking, along with biometric trackers that monitor pupil dilation, mood, and other biometrics will make immersive media the most "personalizable" medium for advertising, ever.

Certainly, an immense market opportunity lies here, but the ethical questions loom even larger. As a child, ice cream was a common cure for a bad day at school or losing a Little League game. So what is the harm in an advertiser using your present mood data to sell you ice cream if you seem a little down? That seems innocent enough. But what if they also know that you're on a crash diet, or you're exhibiting signs of depression because you continue to fail at your weight loss goals? Is it okay to offer you ice cream then? How about alcohol or prescription drugs? Or a gun? Certainly, individuals must have the wherewithal and responsibility to make good decisions about their health and livelihood, right? Right?!

One can imagine similar scenarios ranging from very helpful to incredibly destructive at every conceivable scale.

Luckily, the Spatial Web enables users to maintain a sovereign ID that can securely store and approve which advertisers have access to which information. This information can be conditionally set by location, time, mood, and/or buying cycle. This is actually good news for advertisers because they have a hard time trying to target users with the right ad at the right time. This is because they purchase a patchwork quilt of data from a number of different providers in an attempt to target you. However, in many cases, they are being sold bad or fake data. Often they are targeting users that don't even exist; they are just bots, or they have no idea of where you are in a purchasing cycle, so they throw money away chasing you around the web with a product you looked at a week ago and bought somewhere else.

Privacy laws and regulations like the European Union's General Data Protection Regulation (GDPR) standards and the California Consumer Privacy Act of 2018 are landmark regulations that are imposing serious fines on tech companies that do not properly handle personal data or do not allow users to remove themselves from their services. But advertisers would be far better served by access to accurate data provided by users directly—data that is correct, validated, up to date. And much of this user-provided data can easily be managed by a personal AI.

How Advertising becomes Commerce

The greatest irony about advertising in the Web 3.0 era may be its likelihood of shifting from hyper-contextual advertising to hyper-contextual *commerce*. This is because the Spatial Web allows any person, place, or thing to have its own digital wallet and to transact in digital currencies and even in micro-payments. How does this shift happen? The point where you might encounter an ad in the physical world or in a virtual one can just as easily be a point-of-sale for that digital good, product, or experience.

For example, imagine that you are in the middle of future Times Square in NYC. You have the latest AR hardware on. You access the Spatial Web through a universal Spatial Browser. Based on the personal data profile that you've granted to any advertiser within range, you would see a personalized set of billboard and holographic ads. Advertisers and products that you are explicitly NOT interested in seeing would not even appear. Your personal AIs would block from view any ads or products that aren't within the predefined range of your tastes or interests. This scene would look similar to scenes in movies like *Blade Runner* or *Ghost in the Shell*, although all of the content and objects would be completely personalized to your taste preferences.

Now imagine an ad appears on a billboard in front of you offering a new set of 3D holophonic AirPods. You can view them floating above you or simply motion for them to fly into your hands and see them in actual size, rotating the virtual AirPods as if they were physically present. You could modify the color, features, and materials before selecting your choice. At that point, you could merely signal verbally, biometrically, or otherwise and your digital wallet would make a peer-to-peer payment once your AI assistant had validated the authenticity and history of the vendor. You can have the product 3D printed on-demand at home so it will be ready for you when you arrive. Or you can have the product transported directly to you anywhere along your path via car or drone within minutes.

Now imagine instead of a pair of AirPods, the ad is for a new Flying Tesla Model Z. In this case, you could have the virtual car lowered to the ground and virtually step into it, experience a test flight through Manhattan as if it were real, and then purchase the actual vehicle with similar delivery options. The exchange becomes even easier if you are being offered a virtual car from another user, say a vintage one-of-a-kind 1981 Camaro Firebird for one of your avatars in the '80s virtual world that you frequent. In this case, the user profile that you allow advertisers to access isn't limited just to your physical self but includes certain avatars that you have enabled with "cross-world" advertising. Just like the "retargeting ads" that you see today—the ones that seem to follow you across the web—these "cross-world" ads can do the same except that they can move between virtual worlds and the physical world to offer you personalized, discounted ads for virtual goods.

In this virtual vintage car example, whether you choose to purchase while physically in Times Square or opted to do so when you saw the same ad appear later in a Tokyo 2100 virtual world as a 3D hologram on the streets, you could simply authorize payment and put the car into your Smart Account inventory along with the rest of your personal assets or have it ported to the bitchin' garage in your '80s World home, or to any other location for which you have transport rights.

Considering this scenario, the collapse of the advertisement and the transaction seems inevitable. This could allow entirely new categories of monetization beyond online advertising and traditional e-commerce to emerge. But for this to work, the Web needs a commercialization layer upgrade.

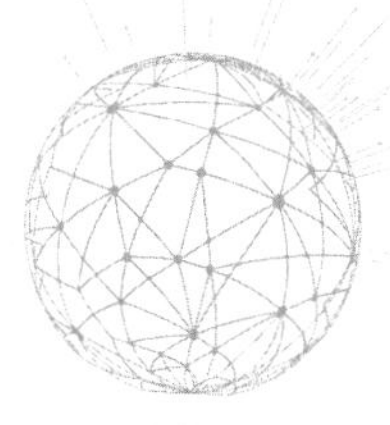

SPATIAL ECONOMICS

From its earliest beginnings in the 1960s, the Internet was born from a set of fundamental principles based on openness, inclusivity, collaboration, transparency, and decentralization. These principles were then embodied in a set of open standards and protocols that we still use today. The Internet is a Decentralization Engine, not only technologically but socially, politically, and now... *economically.*

While its original premise was the voluntary exchange of data across a decentralized network of networks with no central authority, its social, technological, economic, and political impact has been profound. Its impact continues to increase every year as it pushes the envelope of the potential of decentralization via technology into every area of human life.

But to enable secure and interoperable transactions in Web 3.0, new tools must be made available.

Interoperable Web Wallet

The Web has no integrated digital payment solution for the use or sale of digital and physical items: no universal secure "web wallet" or native digital currency with the ability to facilitate peer-to-peer microtransactions and provide for near real-time global payments and settlement. Authentication, tradeability, and transactability of valuable virtual assets (which in some cases will need to be authorized by pupil, voice, or gesture) will require the support of Distributed Ledgers. IoT devices will trade real-time resources and data. AI will manage international finance. Siloed e-commerce carts of today's web will simply not suffice.

The web wallet of the Spatial Web should be integrated into a Spatial Browser as part of your Personal Smart Account. It should be able to handle transactions with any currency or payment service—fiat, credit card, cryptocurrency, or other. It is possible that a universally accepted native web currency could emerge. Many people feel that this is the promise of Bitcoin. Whether this is the case or another currency assumes this mantle remains to be seen. In any case, a native web wallet built into a Spatial Browser as part of a single-sign-on that works interoperably everywhere would open the possibility for entirely new categories of economics—it connects the human, machine, and virtual economies to come.

The Economy of Everything

By definition, the Internet is a technical system. However, over the past two decades, the Internet has become far more than just a technology. It's a decentralized communications infrastructure that enables networks around the world to interconnect without needing the approval of a central authority. With more than 3.5 billion people online today, the Internet is now becoming a decentralized economic entity larger than the GDP of the UK, India or Brazil. And it will evolve substantially over the next 10 years, fueled by innovations in technology and business models. The Spatial Web is the key to enabling a Digital Network Economy powered by the exponential technologies of AR, VR, AI, IoT and DLT's of the Convergence.

In our hyper-connected global economy, no sector of the existing economy will be untouched by these technologies—hospitals, farms, cities, manufacturing firms, transportation companies, retail stores, media, gaming, and virtual worlds will be impacted and transformed—and only those who adapt quickly to technological change will be successful. Business models and the very nature of economics may be profoundly changed.

Think of it as the Economy of Things, the Economy of People, the Economy of Places, and the Economy of Experiences all being networked together. This will be brought about by the creation and then networking of four major trends.

Digitization of people, places, and things
Transactionalization of physical or digital places and things
Spatialization of transactions
Monetization of novel sensory experiences

Spatial Retail Shopping

Amazon Go's new cashierless stores are shaking the foundations of traditional retail by automating the entire shopping experience as a frictionless transactional environment. Automated retail has been around for a long time. We've all used it. Vending machines have provided us with sodas and snacks for decades, and if you're in Tokyo, you can even purchase apparel from such machines.

However, in the Spatial Web, you will be able to walk into any store, pick up what you want to buy, and be automatically charged for your purchases as you exit. No need to wait in line.

The "just walk out" future of shopping allows you to enter a store by checking in with a smartphone, and with identification and verification or by cameras with facial recognition software. You can browse and add items to a basket as in any grocery store. Every time you pick something up, add it to your basket or put it back, a combination of

AI, computer vision, and sensors track your movements. When you are finished shopping you simply walk out of the store and are automatically charged for the items you took…but you can see how quickly this becomes problematic in a siloed, "app-specific" ecosystem.

Will only Amazon devices work in Whole Foods or Apple devices in some stores, Google's or Samsung's or Baidu's in others? What about in a mall or shopping district? How will this work if I need a separate app account and wallet for every device, operating system, or store? The Spatial Web must enable interoperable and automated retail experiences.

Imagine the "Amazon Go-ification" of every store, just not exclusively by Amazon, but as part of an open standard that any storefront can use, every device can reference, and any wallet be used to transact. For that, you need the benefit of a shared data layer—a single source of truth that can be referenced by anyone with the right permissions for Identity, Assets or Goods, and Payments. In other words, you need a Single Sign-On for the World that comes with its own Web Wallet.

Perhaps even more interestingly, a new sector of the global economy will become the fastest growing and ultimately the largest sector— the Experience Economy.

The Immersive Experience Economy

As VR and AR begin to generate high-quality digital experiences in an endless number of virtual environments, new generations of consumers will increasingly choose to purchase experiences over services and products. Older jobs will be automated and the Experience Economy will become the largest sector of the economy. The term "Experience Economy" was coined by B. Joseph Pine II and James H. Gilmore in 1998. They described the experience economy as the next economy following the agrarian economy, the industrial economy, and the most recent service economy.

Services today are increasingly being commoditized, is that because of technology, increasing competition, and the increasing expectations of consumers. This situation represents one of the strongest arguments for an economy based on experiences, especially ones that will predominantly be virtual, immersive, and personalized.

Historically, products can be placed on a continuum from undifferentiated (referred to as commodities) to highly differentiated. Just as service markets build on goods markets, which in turn build on commodity markets, so experience markets (and eventually transformation markets) build on these newly commoditized services (e.g., internet bandwidth, consulting services, etc.).

The classification for each stage in the evolution of products is:
- A commodity business charges for undifferentiated products.
- A goods business charges for distinctive, tangible things.
- A service business charges for the activities it performs.
- An experience business charges for the feeling customers get by engaging it.
- A transformation business charges for the benefit customers (or "guests") receive by spending time there.

You can see the forerunners of this in the gaming world, in social media with Instagram and Facebook videos—and increasingly Snapchat lenses, and with the introduction of volumetric video that creates 3D holograms much like R2D2's projection of Princess Leia in *Star Wars*.

In 2018 Uber, Audi, and Disney introduced synchronized VR "Holorides" that enable users to "skin" or add a layer over their drive that can stylize the road and scenery outside their window to look like a cartoon or a movie. In a Holoride, you can play a game of Angry Birds along your route or battle digital characters in a simulation of Tron that is mapped onto the road itself. Imagine the range of experiences enhancing and overlaying every inch of the real world itself, magnified by the potential billions of virtual worlds and experiences.

In 1996, Danish researcher Rolf Jensen of the Copenhagen Institute for Futures Studies wrote in his article "The Dream Society" for *The Futurist* that American society is yielding to a society focused on dreams, adventure, spirituality, and feelings where the story that shapes feelings about a product will become a large part of what people buy when they buy the product. Jensen framed this trend as the commercialization of emotions. "In 25 years, what people buy will be mostly stories, legends, emotion, and lifestyle."

The first eCommerce transaction occurred in 1994. It launched the eCommerce Era. Twenty years later eCommerce drove $1.5T in B2C transactions. In 2019, China alone surpassed $1T in eCommerce transactions.

All this has happened before even factoring in an emerging machine-to-machine (M2M) economy where the smart, autonomous, networked, and economically independent machines act as the buyers and sellers and service providers, powered by AI and able to carry out significant physical and economic activities with little to no human intervention. This evolving ecosystem will be made possible by the growing number of IoT devices able to trade and share compute power, storage, and data.

Spatial Economics will exponentially increase our global economic value as it combines our physical economy, the machine economy of IoT, and the digital economies of billions of augmented and virtual environments, their assets, and AI agents.

This presents us with another critical question—how will we search and navigate across something as vast as the Spatial Web?

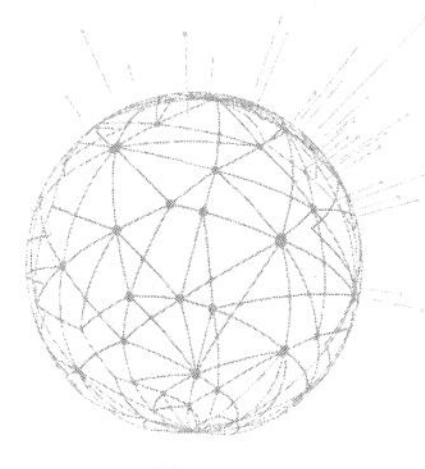

NAVIGATING THE SPATIAL WEB

Searching the Spatial Web

Search as we think of it today will change radically as smart glasses become commonplace. Today, we predominantly search the web via text. We type in a query. Then Google searches the content it has indexed from across the web and delivers to us the pages it considers to be the most relevant based on an algorithmic page ranking model. You can't search for or discover assets, objects, or users while you are at a specific location, because Google has no index for those things. On the Spatial Web, *what* we'll search for and *how* we'll search for it will fundamentally change.

In a Spatial Web search, your personal AI will check spatial contracts using a combination of object and sound recognition to detect how and what is presently in the location you are searching. Spatial Search will react to your voice commands, track your eye movements, gestures, and emotional and neural responses, filter for your historical taste preferences, select the optimal spatial content channels to display, and feed all of this into your Spatial Browser. This will let you find what you're looking for, thinking about, or desiring, and present it to you based on an increasingly subtle and nuanced set of personalized results—tailor-made for you—presenting the right thing that you are searching for and making it available, exactly where and when you want it.

You can search via domain name, location, asset, or channel. For example, you might say "show me the city streets" or "show restaurants" and the Spatial Browser or application would perform the search, enabling relevant Smart Spaces and Assets in your vicinity to appear, and loading any Smart Contracts, with the appropriate permissions required to view or interact with them. Smart Spaces will also have default information associated with them, stored with their asset content, including Channels, keywords, and arbitrary text that will be crawlable.

If you're sitting at home on the couch with your VR Headset, you might want to go shopping. When you say "Let's buy some shoes," your Spatial Browser might know from your preferences to shop at Foot Locker, so by default, it takes you to the Foot Locker Space. Or perhaps you are at home using AR glasses and wish to search for shoes across multiple stores. The AR glasses can give the stores access to your size, shape, and color preference ranges, allowing you to simply swipe through them, project them onto your feet, and confirm before purchasing via your secure wallet plugin to your browser. You can have the physical shoes droned to you within an hour and the virtual pair delivered instantly to wear in a game of virtual basketball with your friends.

Inside of a physical store, the Spatial Browser will be smart enough to perform local searches, displaying menus of available content. Spaces can have their own channels that offer complete flexibility for making content in any Space discoverable by users. So you could search inside the shop for heels, then pick a pair from the results, look at them in detail, or wander the virtual shop floor to browse the displays.

In a physical grocery store, looking for a specific product, you could ask "Where is the coffee and based on my previous coffee purchases, which organic fair-trade roaster would I like?" The Browser will perform a localized channel search of the grocery store's Smart Space, find matching products and direct you to the correct aisle and shelf via AR directions.

Content can be searched for by channel globally, within a Space or even for a user. On the Spatial Web, users can add content or Assets to a channel in your User Inventory Smart Space which can function as a micro-feed like a combination of Twitter, Instagram, Facebook, and Blogging /Vlogging for that user. Other users can subscribe to the feed by subscribing to that channel. Enabling a "UserFeed" channel globally would allow any content across all feeds to be searchable by tags.

Perception in the Spatial Web

The ability to perceive (see, hear, touch, etc.) anything on the Spatial Web is determined by the permissions as defined by the Spatial Domain where a given Asset is located. Merely being able to search for something does not automatically grant the right to see the information, object, or person. If the Smart Space that an Asset is currently in is searchable, the Asset itself has to agree that based on your user permissions, you have the right to even perceive it. Beyond that, perception is not limited to an individual's specific biological or mechanical senses as any combination of human, machine and virtual sensors and haptics could be combined to experience or analyze something. Both human and computer vision combined along with several IoT devices that measure temperature and moisture could be required to validate that a particular medical device is approved for surgery. In the Spatial Web, perception can be augmented, modified, shared or crowdsourced.

Interaction in the Spatial Web

Interaction in the Spatial Web can be governed by Spatial Contracts that determine the interactions of any user with any object or other

users in any space. This may include creating an asset or particular types of interactions including but not limited to touching, moving, routing, rotating, scaling, modifying, consuming, or destroying, all of which can be governed by rules, rights, and governance layers based on the rights granted and recognized by the authorities of a given domain.
Transacting on the Spatial Web

Transactions in the Spatial Web can occur between any person, asset, or space under any conditions set by Spatial Contracts. Transactions can occur automatically—the terms can be preset within a Spatial Contract to initiate a transaction between any two (or more) parties, in any accepted currency, using any form of virtualized wallet.

The invention of Bitcoin has been compared to the invention of email. At first, email was only used by computer experts to communicate on the internet worldwide, instantaneously with anyone who also had an email account. Everybody else continued to use the postal system for another 20 years until personal computing became widespread.

Bitcoin can be thought of as eMoney. It can be sent over networks worldwide instantaneously whereas our existing national currencies have to travel through multiple intermediaries to arrive at their destination. This is both time consuming and costly.

Today, everyone uses email, text messaging, video conferencing, and other digital communications tools. Rarely do we mail a letter. Tomorrow, everyone will use various forms of digital money and rarely will they use national currencies. This will fuel the next great wave of digital commerce and it will enable many forms of automated transactions between spaces, devices, and objects with no human interaction required.

Transportation on the Spatial Web

The transportation of Assets or Users in the Spatial Web occurs between Spatial Domains. This can be a function of "hyperporting" any digital asset between any Smart Space across geo and virtual locations or dynamically routing physical objects along a path that includes any number of points in three-dimensional space. For example, you could easily hyperport between any number of virtual worlds as if they were all a part of the same universe, just like "moving" between websites today. Similarly, moving from point A to B in the physical world would simply be a matter of routing a user between two Smart Spaces, much like mail delivery or an Uber car routes packages and people between physical world addresses today. And since Smart Spaces exist in the physical world as well as virtual worlds, one can imagine a one of a kind BMW concept car co-designed by multiple parties from around the world using VR. Upon completion, the virtual car could be teleported to a viewing room in London for an audience to see in AR. We call this cross-reality transportation.

Time in the Spatial Web

In the Spatial Web, you can specify a point or a range of date and time you wish to view. When present at a particular location (physically or virtually), you can "scrub" through a timeline to go back in time to view the contents of the space at a given time (if the data is available and you have permission to access it), including full historical representations or information at a given place. Conversely, you could place objects to be viewed in the future or to be used to view the timeline for a future project.

Channels in the Spatial Web

One of the most significant challenges to AR and VR is managing and filtering the available content or permissible activities of a given space within a given category. Imagine an unfiltered AR Times Square experience, with thousands of sources of content and advertisers; the experience would be completely overwhelming. Enabling "channels" of content that allow users to select the content they want to view enables a personalized AI to filter content and activity based on the permissions and preferences of the user. For example, the user may only be interested in seeing Times Square restaurant reviews on Yelp, Instagram photos from friends, virtual objects from an AR game, and Local Municipal content. In addition, users can create their own private channels where they place their own content for personal use or to share.

When immersed in an AR or VR world, people's behaviors will be significantly different from how they use their smartphone and desktop interfaces today. Currently, when using a smartphone, people enter into discrete applications for specific functions, usually one at a time like posting a photo to Instagram, sending an SMS, or making a phone call.

In AR or VR, the device is merely a window into a world where multiple layers of content can be overlayed on top of one another. Reality itself then becomes both the actual experience and also simultaneously, the interface for interacting. For example, you may be able to see messages from friends, entertainment published by favorite artists, instructions for a work assignment, all viewable at the same time by any filter or query, similar to tabs on a browser.

As the value of these new digital layers of content and data increases, the quantity will grow exponentially, especially in high traffic and complex areas such as city centers, industrial areas, and transportation hubs. In order for people to interact naturally and seamlessly in the digital world, the relevant content needs to be made available at the right time for the right purpose, and in the proper context. Managing, filtering, and securing these layers of content poses one of the biggest challenges to the Spatial Web—we address the problem with what we call *Channels*.

Channels are worldwide layers of content or interaction from a specific source or organization. For example, Facebook, Instagram, and Snapchat can all have worldwide channels that can be seen simultaneously, enabling users to see the content posted, based on *where* it was published rather than a list of posts in the feed based on the *time* published. Any person, organization, or municipality can publish content into the world, and so technically there could be an unlimited number of potential channels. The right to access the channel is up to the publisher who sets the conditions for who can view, interact, post, edit, and transact. Similar to web browsers, channels may come with plugins that enable new engagement possibilities, allowing the platform to naturally evolve and innovate.

When a user enters a space, the AR/VR browser searches for all available channels and then, based on that user's preferences, permissions, and privacy settings around what they want to see proceeds to render the relevant content across the correct channels simultaneously. For example, a user might specify "only show me social posts from friends" or "only show me work-related content between 9am-5pm or while I'm at work" or "show me all fire hydrants."

As time goes on and the number of channels increases, the filtering of content will become critical. Imagine walking into Times Square in the year 2040. The layers of content, built up over years of posts and content publishing, will be impossible to view all at once and overwhelming even if you could. As the amount of available content grows over time, manually managing preferences will become more daunting and will likely result in missing relevant chunks of information. A Personal AI Agent will easily be able to manage this and tailor the user's experience as it learns what kinds of information and content they like. And especially critical, the Personal AI Agent will handle data in a manner that does not require explicitly sharing any personal information with any other party.

As a market for personal AI agents develops, each promising to deliver the perfect custom experience for the user, users may choose to frequently adjust or swap AI assistants based on mood or situation. As these AIs become more advanced, curated AIs will offer a whole new media type created by magazines, cable channels, thought leaders, or tastemakers. Imagine having *VICE Magazine*, The History Channel, and Malcolm Gladwell curating AIs all running at the same time, providing a completely new and immersive way to learn, share, and express. While walking through Tokyo, you may be pointed to an obscure Bansky piece, while viewing the world through a seventeenth-century historical Japan lens, all while receiving an audio-visual social commentary from Malcolm Gladwell.

Spatial Domain management will protect businesses and private property from becoming virtual billboards for would-be advertisers by enabling their owners to select what if any content can be posted in their location and what rental fees may apply.

Some social channels may find new ways to "prune" spatial content that would otherwise be permanent. For example, imagine the future Snapchat of the Spatial Web; the channel may quickly become filled with an overload of creative content and cute AR creatures. By making the content ephemeral but likable by viewers, content that gets "likes" from passersby gains additional life, perhaps even becoming immortal if enough people like it, while less admired work fades away and is only viewable to the poster and perhaps their closest friends.

Such a rich and immersive world may not be ideal for all situations and environments, however. Some places such as the workplace may become "digitally hygienic" environments, allowing only work-related content to appear, or a school may not allow commercial game experiences during classes. Furthermore, depending on who you are, certain channels may or may not be available to you. For example, doctors could have access to channel information that medical students might not, while the building maintenance crew could have their own dedicated channels. The conditions for these restrictions will be set by the relevant authorities.

Channels don't necessarily need to be worldwide. Often they may only be useful in specific domains. For example, a restaurant may only provide content within the walls of its business or a hospital may only publish sensitive information within the property. Certain domains may require people entering into their space to view a certain channel, overriding their preferences, such as when you walk into an airport, they may force you to view a public message, or a restaurant may automatically present their menu channel (which could then be opted out perhaps for their next visit), or the hospital may force patients to view the hospital channel with no option to opt-out.

With this new world of virtual content, channels and spatial rights, no doubt abuses will occur and people's rights may be infringed upon. The Spatial Web architecture is designed to support that fight.

Security on the Spatial Web

Smart Spaces are a unique invention in part due to the security capabilities that are built into them. Because Web 3.0 presents many new challenges to security, privacy, and verifiability, new innovations are required. The Spatial Index uses various security mechanisms to control the usage, transfer of Assets, and interactions with a Space and its contents. Spatial Domains are registered on a Distributed Ledger that uses multi-factor authentication and validation. All interactions via the Spatial Browser use multiple layers of encryption backed by Self Sovereign Identity ledgers and multi-factor biometrics. Every Smart Space gets its own Channel by default. Registering content and Assets to a secure channel enables the system to control which users and Spatial Browsers are able to detect them.

All transactions in and across the Spatial Web occur via encrypted protocols, and sensitive data can be stored on private ledgers or encrypted on a public ledger. Spatial Contracts allow additional security for Spaces and Assets, controlling when, how, and by whom transactions can be made across the Space or against individual assets.

The nature of decentralized digital identity automatically protects users of a Space from data mining, preventing the correlation by outside parties of activity within the space to any other activity.

Spatial Browser

The Spatial Browser is the universal window of the Spatial Web. It allows Spatial applications to be installed like browser extensions; they can self-install and self-execute as users enter spaces based on the permissions set out in Spatial Contracts. Although developers can design their own independent Spatial Applications for use across any VR- or AR-enabled operating system, the Spatial Browser will be the primary interface for the Spatial Web, displaying both 3D and 2D objects, environments, and animations. Using "Channels," an infinite number of views can be presented and yet be filtered and permissioned by AI to be shared by a group or personalized for an individual. Think of Channels as layers of digital tracing paper that allow any application or service provider on the Spatial Web, whether large like Snap or Yelp or small developers, to function like browser tabs that can be overlayed simultaneously in the world.

The Spatial Browser can have a default public data channel, which automatically renders local street and business data, from maps and directions to parking space information, names and information on stores or restaurants, and even local events. The addition of custom plugins would enable all kinds of custom skins, filters, audio, gesture, voice, and thought-based interactions.

The Spatial Browser is intended to be released as an open-source project for anyone to build on and extend. The Spatial Protocol specification is designed to become the universal standard able to be adopted by any spatial browser, just as the various web browsers of today all share the same underlying web protocol and programming standards. The global developer community and standards organizations must be engaged in the ongoing development of standards for all Spatial Browser implementations with a special focus on standards for privacy, security, interoperability, and digital payments.

RISKS & THREATS

> "Alone we can do so little.
> Together we can do so much."
>
> *Helen Keller*

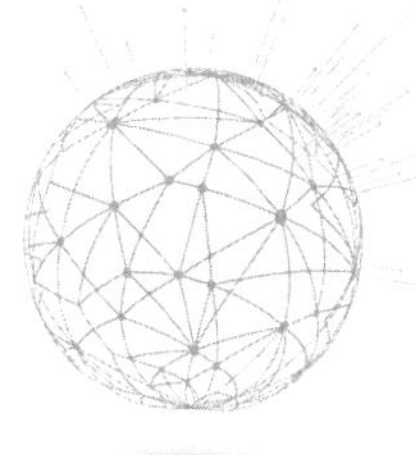

THREATS TO THE SPATIAL WEB

A Spatial Web that is open to all and fully integrated into a single global network of networks—just like the current World Wide Web—would be the ideal. However, tremendous obstacles to realizing that vision will arise.

Although this book attempts to make a strong case that the Spatial Web is an effect of evolutionary trends that make it appear as inevitable, it is, in fact, far from being a certainty. If anything, it will take a significant amount of dialogue, development, and commitment from engineers, creators, thought leaders, non-profits, standards groups, and governments. To assist in creating awareness, advocacy, and adoption of Spatial Web standards, we the authors and our associates have established the VERSES Foundation, a non-profit organization dedicated to delivering the protocols and specifications needed for an open, free, and secure Spatial Web.

Corporate Interests

Many factors can threaten the realization of the Spatial Web. If powerful corporations like Apple, Google, Facebook, Samsung, Baidu, and others believe that they must maintain closed hardware and software ecosystems with respect to emerging technologies like AR, VR, IoT, and AI, and if they are unwilling to participate in adopting open standards that enable interoperability, then this will slow the adoption of new open standards. They may choose this position because they fail to see the benefit, or fear it may threaten their profitability, or they simply are not interested in the idea or the implementation approach. This would result in a series of balkanized spatial webs that are not interoperable. If this happens then huge numbers of the users will not be able to participate in an open Spatial Web.

Even worse, if each of these providers attempts to create their own version of a spatial protocol and spatial browser to make their own siloed spatial web, then users will be forced to choose between them, as the likelihood of any one of them developing their own proprietary commercial standards that their competitors will adopt is effectively zero.

The Historical Failure of Walled Gardens

History has seen previous attempts to wall in open services, usually with poor results for all involved. A "walled garden" is a network or service that either restricts or makes it difficult for users to access applications or content from external sources. Standard TV and radio are open—anyone with a television or radio can access them; cable TV and satellite radio are walled gardens, requiring users to subscribe to view channels and programs. In the early days of the World Wide Web, companies like America Online (AOL), Prodigy, and CompuServe functioned as closed and curated versions of the Web. They only made available the websites of affiliate partners for their paid subscribers. Although these companies drove meaningful early Web adoption, users were eventually driven to climb over the walls and make their way into the open Web itself.

The Music Industry offers a cautionary tale about the long-term effectiveness of walled gardens and closed systems. In the year 2000, annual worldwide music revenues peaked at about $50 billion. This was due in large part to the digitization of music and the introduction and growth of the compact disc (CD) format in the 1980s and '90s. In 2000, Napster emerged from the shadows and allowed anyone with an Internet connection and little concern for copyright laws to download unlimited amounts of music for free. The record labels freaked out, sued everybody, and lobbied Congress to stop the piracy. Apple finally stepped in to offer a solution in the form of the 99-cent song, protected by Apple's closed ecosystem and a security protocol called Digital Rights Management (DRM) that locked in their proprietary audio format on Apple's iPod music player.

Between 2004 and 2009 when DRM was being enforced, Music Industry revenue in the US dropped by nearly half. Market research company NPD reported that same year that only 37 percent of music acquired by US consumers was paid for. Piracy had actually increased during this time, and in the end, the labels were forced to open up the content and adopt an open audio format.

Ultimately, this mattered little, as the era of streaming music would force the industry through another format change and down a new path. Music streaming aggregators like Spotify, Apple Music, Tidal, and Deezer were able to license nearly all of the music available for users to access via paid subscriptions. At least with music, the vast majority of these services have a similar music catalog, opting to compete with various filters and personalization features as the differentiators for users to choose between. This is, at least, an honest approach to the industry that respects the user and their desire to access a broad catalog. In the rare cases where certain artists restricted their music to a single service that they were more aligned with politically or monetarily, the fan blowback was so severe that they eventually gave in and released the material across the rest of the services.

According to the International Federation of the Phonographic Industry (IFPI) 2018 Global Music Report, the greatest irony in the case of the music industry may be that the largest music streaming service for the last decade, with more than double the amount of listens than the rest of the other music services, has been a VIDEO streaming service…YouTube. The open and unrestricted access to music, even via a video service with ads, remains preferable to users over content-specific services. One has to wonder if it is because users today are multi-content, multi-mode, and multi-device. Switching between YouTube videos and music seamlessly is probably preferable to opening another app. Furthermore, a Youtube link is a standard web url and therefore a universal sharing format that works for everyone. The Music Industry still makes it extremely difficult to share a song file. Think about it. How many Youtube links have you shared versus the number of song links from an app? Walled gardens crumble when network effects hit them.

The paid streaming video apps of today like Netflix, Hulu, HBO Go and Amazon Prime are also walled gardens that restrict their original branded content to their own apps. This is causing further fragmentation in the on-demand video streaming market. In 2018, Disney announced its plan to pull its content from the others to exclusively offer it via their own Disney Plus subscription app. So, what began as a new open format to view premium movies and television shows over the Internet is quickly becoming a series of separate islands that restrict consumer choice and force users to adopt multiple apps and services to access the content that they want. While companies have the right to license and monetize their content in whatever ways that they see fit, it is easy to see why consumers eventually retaliate and start scaling the walls.

The IOS and Android development environments and app stores are also types of walled gardens. These separate platforms require developers to build and maintain two different versions of their apps if they hope to reach the largest number of users. This is not an insignificant challenge and adds additional development costs that small developers may not be able to manage. This can have a chilling effect on innovation which can limit user choice to larger app developers able to bear the costs. And it becomes a nightmare for consumers when they need to switch between devices, as the process of locating, downloading, and logging in to their apps on a new operating system is an exceedingly frustrating experience. But, these are Web 2.0 problems. In the Web 3.0 era, walled gardens become a big threat to an open Spatial Web with the arrival of AR or Smart Glasses.

An Un-Interoperable World

In many ways, practical requirements for the Smart Cities of tomorrow make them the ideal showcase for emphasizing the need for the Spatial Web.

Imagine yourself navigating through a retail district in a decade or so in your city. You are wearing a pair of Amazon smart glasses, PrimeLens. Everywhere you look you can see the content and holograms of the partners and applications that are part of Amazon's spatial web. As you make your way into a Whole Foods (owned by Amazon), your glasses identify you and "check you in" to that location. Personalized offers appear for you with special discounts. Your grocery list appears and you simply follow the arrows to each item. You grab the items that you want and you're automatically charged as you exit. However, if you go next door to Foot Locker, an Apple partner, you'll need Apple's iGlasses to access their spatial content, special offers, and auto-checkout.

Here's another example: a husband is at Macy's, shopping for his wife but he can't get access to his wife's body size digital avatar because he uses Microsoft's Hololens for work, and she can't send the avatar to him without checking into Macy's location first. Beyond this type of "bad mall" shopping experience, how exactly will Smart City content created for all citizens be accessed? How will municipal content such as street signs, parking places, and other information placed by the city work? Will it be stuck in spatial municipal or corporate siloes too? Without agreed upon standards, this is a recipe for chaos for civilian uses and an administrative nightmare for the Smart City staff and workers.

Even worse, how exactly will multiple parties view and interact with the same content at the office, at home, in a park? Will companies need to standardize on a single device for every employee? Will everyone need to download the same app in order to look at an architectural history of the historical buildings, play a shootout game in the living room with family, or chase and collect new virtual characters in the park?

How exactly will a Smart City of the future work if none of the so-called "smart things"—smart glasses, smart cars, smart stores, and smart accounts—work together...How *smart* would such a city be?

You can see how the combination of spatial content, centralized data storage, and proprietary payment systems are not only impractical for our future Smart Cities, but will be entirely dysfunctional for everyone, no matter where we go.

You can make walled gardens for individual users staring at a screen by themselves. But you can't *silo* the world. You can't succeed in making a closed "Spatial Web" any more than AOL could make a closed web 25 years ago. The web needed to be "world wide." But there is really no need to have these silos.—there is plenty of room in the Spatial Web

The open specifications and standards of the World Wide Web enabled many of the Web 2.0 companies to drive themselves to trillion-dollar market caps. Android and iOS are constructed from the open-source benefits of Linux. For Web 3.0, there is no need for companies to compete at the level of the core spatial protocol or kernel—they will all be free to develop their applications on the Spatial Web. The best thing that large corporations can do is assist in the support of open Spatial Web standards and the Spatial Protocol that is designed to connect us all. Then we can all benefit and thrive from the network effects that we achieve together.

Government Bureaucracy

Another threat to the Spatial Web is government confusion, opposition, or apathy regarding the evolution of privacy rights, property rights, and legal frameworks for digital currencies. We will need technologically savvy parties with strong government and public sector relationships and backgrounds to assist in the education process necessary to gain widespread support for the effective adoption and implementation of the Spatial Web. This is especially true in light of the legal and regulatory implications that the Spatial Web poses and the need to re-frame and up-date our definitions of property rights as it relates to private, public, and intellectual property use regarding AR content and its adoption at scale.

In addition, the ability for the Spatial Web to revolutionize our global economy and serve as the economic engine that powers wealth creation for the next century will be largely determined by the implementation of common-sense regulations of digital currencies and digital assets that strive to protect users without constricting innovation.

The effective implementation of a seamless and frictionless peer-to-peer global economy that unites the human, machine, and virtual economies together could deliver a level of prosperity that could (finally) enable equitable economic benefits for all of the inhabitants of the planet.

This is an opportunity far too great to allow potential threats and risks to the Spatial Web to prevent us from working together, with all of our collective will and power, to make it a reality.

IMPLICATIONS

"In all human affairs there are efforts,
and there are results, and the strength
of the effort is the measure of the result."

James Allen

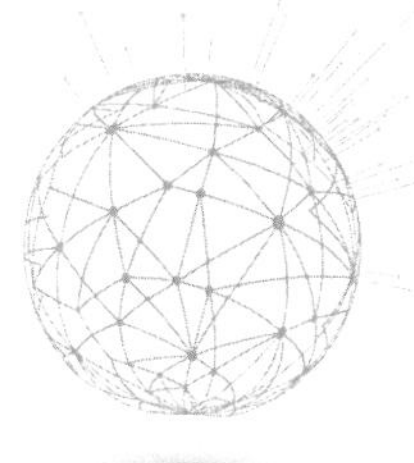

FROM THE
SPATIAL WEB TO
A SMART WORLD

There are an infinite number of applications of the Spatial Web. Fundamentally, the millions of applications that it enables are the result of enabling all "smart things" to work together like smart glasses and cars and factories and cities, smart payments and contracts and assets and identities and spaces. These Smart Spaces, all networked together, hint at the Spatial Web's largest implication. For the first time in human history, it would enable a smart and interconnected global civilization. ***A Smart World***.

Dawn of A Smart World

A Smart World is a world where there exists a self-sovereign, universal identity and address for any person, place, or thing, across both physical and virtual domains. Smart payments and smart assets are integrated into smart cities. The Spatial Browser, working across all brands of smart glasses and other new interfaces, enables everyone anywhere to access location-based smart information and objects across real and virtual worlds. And finally, a Smart World is governed by a digital ocean of Spatial Contracts that permeate everything around us, allowing dynamic rules to be automatically executed and enforced in order to enable the interactions, transactions, and transportation of assets and users between Smart Spaces.

Cities are busy places, full of roads, vehicles, energy and waste management, resource systems, planning, laws, and bureaucracies. In a Smart World, cities will track, monitor, and optimize the flow of traffic, power, water, waste, resources, goods, information, transactions, and people. They will optimize usage dynamically and make adjustments accordingly. Municipal data relevant to locations can be displayed and updated in real time. Government and decision making can be revolutionized with systems designed to enable citizens to contribute to the decision-making process. Such a Smart World would make all city infrastructures in the world compatible and interoperable—because a Smart City is just one type of Smart Space. And there are many types of Smart Spaces.

Smart Factories are another type of Smart Space. Modern factories have millions of different components and processes with highly complex machines in operation producing a myriad of different products. The modern factory is a very complicated system. But in a Smart Factory every part, robot, device, and product becomes a Smart Asset. All of the materials, machines, and products are tracked through space and time. Smart Assets contain the Smart Twins of each item, which also contain the records of their maintenance history secured by Blockchains. And all of this information is spatially attached and tracked to Smart Assets via AR and Computer Vision all working together. When a factory becomes a Smart Space, it can set the rules to manage, record, and authenticate the movement of all the products and all the activities of both humans and robots. AI and Smart Contracts can be deployed to analyze and execute operational requirements to optimize the ease, efficiency, and effectiveness of the factory. These same capabilities can also be applied to Smart Farms, Smart Mines, Smart Stores, Smart Homes, or Smart Cities.

A Smart City and a Smart World use digital wallets for transactions. But in this new digitally-transformed environment, everything has a wallet. Not only people have wallets, but objects and spaces also can have wallets. This enables anything to transact with anything else. For example, as a Smart Asset enters a Smart Space, such as an autonomous vehicle entering a parking structure, its identity and purpose are authorized. Then as it pulls into the parking structure, a payment is automatically triggered between the car and the building based on the permissions set by the building rights holder.

Smart Transportation

By extension, it is easy to see that this kind of Spatial Commerce forms the basis for an entire unified global smart supply chain. Smart Spaces that are connected to one another, using smart payments, authenticating users, spaces and assets, tracking interactions and transactions between everything, enable global supply chains.

How does a global supply chain work? First, a Smart Space would grant the ability to gather raw materials and transport them to a Smart Space manufacturing plant to turn the raw materials into a product. The product is then assigned its own Smart Asset ID, and based on the rules contained in a Spatial Contract, it can be moved from point A to point B to point C, from manufacturing to shipping and from storage to store with a complete record of what happened if an audit is required. All of the payments and transfers of ownership can all be automated and recorded. And consumers can see whether the product fulfills particular brand promises (e.g., Fair Trade, conflict-free, carbon-neutral, sustainably produced), or not.

Furthermore, a Smart World enables connected spaces that allow users and objects to move between them. Instead of the pages of the web, every space becomes a Web Space, and instead of moving between pages, you can now move between virtual spaces. For example, imagine you are in a fantasy world where you buy a magic sword that would be a Smart Asset with a history of its battles recorded into its Smart (twin) profile. You then teleport yourself in the form of one of your many avatars and the sword to a futuristic world where you battle another user with a lightsaber. If you win the battle, you could actually record that win into the sword itself and then teleport it, with permissions, to your niece or nephew's bedroom as a surprise birthday gift (with their parent's spatial permission). This is just one example of data and object and user portability between virtual worlds and the real world itself allowing assets and people and currencies to move between them. Another way of saying this is that we have created a seamless unity between Augmented Reality and Virtual Reality, an essential capability in the emerging Smart World we are all busy building—whether we realize it or not.

Smart Ecology

The Spatial Web can help humanity to become more sustainable and will be a powerful new framework for Impact Investing. The industrial economy has brought us many benefits but it has also created many unsustainable social and environmental problems. The Spatial Web can help us address these seemingly intractable problems better than any technology yet invented. As a quote attributed to Einstein says, "We can not solve our problems with the same level of thinking that created them."

The Spatial Web enables a new level of thinking. A new way to analyze and solve the massive problems we now face begins with the concept of the Smart Twin. As we continue to digitize our businesses, our cities, our countries, and ultimately, our entire world, we will create a Smart Twin of Earth. We will have an increasingly accurate digital model of the planet as a living system with near real-time measurements of its various systems and the impact they have on us and on our environment.

This model will be online and available for everyone to observe and use—something like a 3D WikiEarth. This Smart Twin will improve the quality of our global conversation because we will have accurate measurements to to inform decisions. No longer will politicians argue over whether something is hot or cold—we will all be able to read the temperature ourselves. With this increasingly accurate model, we will be able to run large-scale complex simulations and do other types of predictive analysis that will guide the way to truly sustainable solutions. Think of it as something like a giant 3D spreadsheet that will allow us to ask a series of "What If" questions and see the potential impacts of our decisions in real-time. Researchers of all types, from anywhere in the world, will be able to use this new tool to help us solve the many problems we collectively face. This is Cybernetic Thinking on a global scale—offering us an entirely new meta-level of thinking about our problems.

The UN has identified 17 Sustainable Development Goals to achieve a more sustainable future for all. The SDGs address the social and environmental challenges that must be solved. They provide a blueprint for how we might realize all of the benefits that exponential technologies can provide us. If we don't address these global challenges, life could become unlivable for large numbers of people.

Utilizing the power of IoT to sense activities, AI to measure and predict outcomes, and AR/VR immersive spaces to learn, experience, and share, the Smart Twin combined with the global collaboration capabilities of the Spatial Web may provide the breakthrough required—

to achieve the scale and number of impact projects that are necessary to make a real difference on such serious issues as climate change, global health, responsible consumption, and all the other Sustainable Development Goals identified by the UN in the window of time that we have to act.

As businesses and governments implement individual spatial computing projects and network them together via the Spatial Web, they will each begin to realize the operational efficiencies of Spatial Operations. Spatial Operations can then utilize AI to provide ongoing Spatial Analysis of these projects. Spatial Analysis inevitably leads to Spatial Optimization which continuously identifies ways of achieving the same or better results using fewer resources. As the Spatial Web spreads around the planet in a manner similar to Web 1.0 and 2.0, we will evolve toward the realization of a Smart Sustainable Planet with an accurate Smart Twin as a powerful real-time visualization, interface and platform.

Looked from the bottom up rather than the top down, the Spatial Web also helps us with specific Impact Projects. As an example of the Spatial Web in action, let's explore the building of a Medical

Clinic in an underserved area of the world. Using the power of the Digital Twin, we can create an entirely new way of designing, building, operating, maintaining, and replicating this Clinic.

Let's begin with the design process. Because the Spatial Web has a Virtual Reality interface, it is now possible for everyone to experience the design of the Clinic as a Smart Twin—at actual scale— no blueprints needed. Anyone can walk through a simulation of the building while it is still being designed—including the various communities that will use the Clinic—and provide designers with feedback. Stakeholder communities will include the medical and administrative staff, the people from the surrounding neighborhoods that will use the Clinic's services, the vendors that will provide equipment and infrastructure, and insurers and regulators. Feedback from these communities can then be incorporated into the design and all stakeholders can sign off on a highly refined final design. This design is then completed by the architects and the final construction document is also a 3D Smart Twin of the building but with every component specified down to the last nut and bolt.

Now comes the fun part. Every component used to construct and operate the clinic can now be bid on by various vendors and the final cost of the building can be locked in. Because the Spatial Web includes blockchain technology in its data layer, all of the building components can be linked to Smart Contracts that pay automatically when the materials reach the job site. This eliminates all corruption and almost all errors.

Using Spatial Web technologies, building the Clinic will now be similar to putting together a child's Lego toy set. Every part of the building process is programmed and sequenced making the building process similar to putting together a kit. Augmented Reality technology is very useful in this stage and you can see a simulation of the finished building sitting exactly where it will be when it's finished. This acts as a visual guide during the building process. The blueprint is no longer on a large sheet of paper but it is right in front of you in full 3D. You can see the building as it will be when it's completed or at any stage in the construction process.

The Smart Twin of the building can be updated in real-time as the building is being completed which is useful for inspection and certification by various city, state, and federal verification teams. This ability to scrub forward and back in time and see every step of the building's construction is called a 4D view—three dimensions of space plus time. As mentioned earlier, once we link and sync a 4D visualization of a Digital Twin with Blockchain data integrity and Spatial Contracts, it becomes a Smart Twin.

In parallel with the construction, it is now possible to begin training staff in the Smart Twin of the building using VR. As the building nears completion, the full staff will have completed training in the various departments of the Clinic, making the handoff from the builder to the owner/operator very smooth.

At this point, we now have two versions of the Clinic—the fully constructed and operational physical building, and a perfect digital copy aka Smart Twin of the Clinic with all of its construction history, materials provenance, legal contracts, and verifiable information embedded within it. Now we can watch the Clinic in action and learn from it. Artificial Intelligence in the form of machine learning can observe the actual use of the Clinic over time and make recommendations on modifications to the operations of the Clinic. This is called Spatial Analytics and provides those with the proper spatial permissions to view the data and allow it to be used to modify their operations, and in some cases, the actual design of the building or of future Clinics that might be constructed elsewhere. Copies of the Clinic can easily be built in multiple locations around the world using the data gathered in the first Clinic to upgrade and refine the design of later iterations.

To look at this process more deeply, the Spatial Web is designed to interconnect all of the exponential technologies such as AI, IoT, AR/VR, Robotics, and Blockchain to enable a feedback loop of Spatial Operations and Spatial Analytics on a planetary scale. This interconnectivity enables the convergence of these powerful technologies to be aligned for good—not just for building more sustainable clinics but sustainable farms, water and waste management, supply chains, and entire cities. They can be used to measure, manage, and coordinate the information and resources necessary to operationalize the global challenges and opportunities outlined in the UN Sustainable Development Goals. Using geo-spatially accurate 3D datasets that can function as a single-source-of-truth that is universally verifiable and shareable, the Spatial Web enables a new level of global coordination that can be actionable across nationalities, political entities, and global businesses for the first time in human history.

The Spatial Web protocol, HSTP, will ultimately create a web out of everything in the world. This will act as a new nervous system for the planet, connecting everything and everyone together. We can now create ever more accurate Smart Twins of our hospitals, our campuses, our factories, our cities, our countries—and ultimately our entire world. Our fundamental human values, when powered by Artificial Intelligence and Blockchain technologies, can optimize energy flows, logistics, and everything else involved in the operations of modern society. This will be used to create and automate sustainable systems and quickly identify non-sustainable activities so we that we can correct or cure many of the global challenges we are currently facing. The Spatial Web enables us to define a clear pathway to finally creating a Smart Sustainable Planet.

A Smart World is the logical emergent result of a Spatial Web that continuously networks together all of these digitally enhanced "smart things"—smart people, smart places, smart assets, smart rules, and smart money—into one holistic system. This digital upgrade for the planet links everything together in an integrated, interconnected digital network economy that enables an entirely new reality.

But what exactly do we mean by the word "reality"?

The Evolution of Reality

What if Humans are not Homo *sapiens* aka the "wise man," a term first coined by the father of modern taxonomy Carl Linnaeus in 1758? What if our "wisdom" or "smarts" is a means, not an end? Although we are intrinsically connected to our biological ancestry including *Homo erectus* ("upright man") and *Homo habilis* ("handy man" or "tool user") as extensions of our bodies, we are not defined by our past alone. <u>It is the result of what we and our tools make that truly defines our species in the modern era</u>. Perhaps it would be more accurate to describe humanity as *Homo realitas*—"the species that makes and remakes its own Reality."

That is the real purpose and function of our tools and technologies. To extend our bodies, senses, brains, and imaginations into the world. To make our ideas shareable in order to work together to improve them so that we can make our reality and hence our lives, easier, more useful, safer, and more enjoyable.

> **"We shape our tools and thereafter, our tools shape us."**
> J.M. Culkin, on Marshall McLuhan, *Saturday Review*, 1967

Consider that ALL technologies are extended technologies. They are designed to transform reality, and in the process, they transform us as well. It is no wonder that many today refer to the combination of our "reality" technologies—Augmented, Mixed, and Virtual Reality—as "XR" or *extended* reality. In evolutionary science, this relationship of animal-tool-environment is referred to as phenotypic expression. It is a co-evolutionary process. We humans create tools to extend our private reality to become a part of a public reality.

The human mind has an utterly amazing skill not found in any other species on earth. We refer to it as the "mind's eye." We can use it to run advanced and complex 3D spatial simulations in our mind. Unfortunately, we have not developed telepathy and cannot share these simulations with each other... yet. We must instead translate these multi-dimensional internal models of the world into simpler mediums and protocols such as spoken language, text, or drawings in order to share them. Consider that most of what we want to share together is lost in translation.

Let's describe this unique skill of creating 3D models and simulations in our mind's eye as our ability to create a "personal virtual reality" or Personal VR. In order to share with others what we are experiencing inside ourselves, we have developed a series of protocols to communicate it. But in order to share, we must reduce and degrade our native spatial understanding of the world. In the process of sharing, our Personal VR loses fidelity, nuance, and context that simply cannot be captured effectively in other mediums.

Language was the first major protocol used for communicating our Personal VR to others. Caves and wall art served as our first ceremonial Public AR theatre, doubling nicely as long-term public data storage for the memories of the whole tribe.

In each Age since, we've used our Personal VR simulators to dream up new and better ways to improve our lives and the lives of others—we've made our reality easier, more useful, safer, and more enjoyable. We've used our tools to alter or augment our reality. Everything that we've ever designed, built, and invented—architecture, products, machines, music, art—is really just Public AR. Our creations translate our Personal VR into a Public AR which becomes a feedback loop that triggers new Private VR simulations that once adapted become new Public AR creations. And this feedback loop becomes an engine of personal and cultural evolution.

"Myths are public dreams, dreams are private myths."
Joseph Campbell

Our various languages and mediums, whether grounded in atoms or flowing through bits, have evolved to increasingly improve our ability to communicate our internal Private VR and convert it to shareable Public AR. What began with sounds and gestures and cave paintings will advance to sounds and gestures and generative immersive digital experiences (modern cave paintings).

The Spatial Web, with the help of advanced Artificial Intelligence, may one day lead to the emergence of a universal 3D digital language. This new visual and sensorimotor language would not be bound to verbal or written words or 2D symbols and shapes. It would emerge as the ideal global, cross-cultural language to better resolve together, our inner and outer worlds. Like telepathy, it could allow us to share more immediately and effectively what we are feeling inside and seeing in our mind's eye. Then we can finally say, "I see what you mean," and it will be accurate, fulfilling one of our deepest desires, which is to communicate and share ourselves with each other, without the fear of miscommunication, in the most intuitive and natural way. This powerful, evolutionary shift in language and understanding could also enable us to explore new territory beyond mere communication—into realms of real-time co-creation of realities that we can play in and inhabit together.

Humanity is a Reality Engine

The tools and technologies that we invented in the Agricultural, Industrial, and Information Ages extended our hands and feet, then other muscles, then our senses, and finally our brains. The key technologies in the Spatial Web stack represent a continuation of the theme of extension of our embodied abilities into the world, just as it has always been.

XR = Input/Output Senses

IoT = Body/Muscles/Cells/ Senses

AI = Brain/Mind

Blockchain/Edge = Memory Storage/Sensory Neurons

Each of these exponential technologies is powerful in its own right. Each alone is capable of transforming our world and our reality in unprecedented ways. But the applications and implications of the convergence of these exponential technologies over the next few decades—especially with the inclusion of Biotech, Nano, and Quantum technologies—presents an opportunity far too important for our global society to ignore. Although we have used our tools to create an amazing world, filled with global communication, commerce, and content sharing, if we are being honest, we know that we have also used them to drive ourselves and the inhabitants of this planet to the very brink of survival.

But there is hope because we are the species that makes and re-makes its own reality, with the most powerful tools in the history of our species. These tools can make reality easier, more useful, safer, and more enjoyable, not just for a small tribe, city-state, or nation, but for the world. With these tools, we can build a Smart World. If we can imagine it in our Private VR, we can create it as a shared Public AR.

The Spatial Web gives us the ultimate medium we need today to project the best ideas that we can envision in our Private VR using our tools (AI and Machines) to collectively share, revise, and manage globally our collective Public AR aka the World. It turns out this is what we do best—*Humanity is a Reality Engine*. The question before us now is this…

With all of the immense power of the convergence of exponential technologies, what kind of Reality will we choose to create?

EPILOGUE

"In the end, we'll all become stories."

Margaret Atwood

"We shall not cease from exploration,
and the **end of all our** exploring
will be to arrive where we started
and know the place for the first time."
T. S. Eliot

THE END